HEAT TRANSFER CALCULATIONS USING FINITE DIFFERENCE EQUATIONS

HEAT TRANSFER CALCULATIONS USING FINITE DIFFERENCE EQUATIONS

DAVID R. CROFT

Department of Mechanical and Production Engineering, Sheffield Polytechnic, Sheffield, England

and

DAVID G. LILLEY

Department of Mechanical Engineering, Concordia University, Montreal, Canada

APPLIED SCIENCE PUBLISHERS LTD
LONDON

APPLIED SCIENCE PUBLISHERS LTD
RIPPLE ROAD, BARKING, ESSEX, ENGLAND

ISBN: 0 85334 720 4

WITH 30 TABLES AND 102 ILLUSTRATIONS

Printed in Great Britain by Galliard (Printers) Ltd, Great Yarmouth, Norfolk

Preface

This book is intended as a comprehensive and illustrated introduction to the use of finite difference computational methods for heat transfer calculations. The methods are basically simple but offer a powerful tool to the engineering designer or researcher faced with heat transfer problems of a difficult, or more often impossible, analytical nature. The text is oriented towards the practical man who needs a complete work to enable him to understand and apply the methods to solve his problems.

The natural starting-point for discussion about this subject is the partial differential equation (PDE) for heat flow by conduction; it is from this equation that the much simpler finite difference equations (FDEs) are derived. The general equation of heat conduction with internal heat generation and time dependence is

$$\text{div}\,(k \text{ grad } T) + H = \rho\, C_{\text{p}} \frac{\partial T}{\partial t}$$

Our study of conduction heat transfer is principally concerned with the solution of this equation, so as to obtain the distribution of temperature or temperature history within solid structures. We shall always be preoccupied with the temperature, because in all cases it is the temperature distribution which determines the rates of heat transfer and the level of thermal stresses.

For certain simple geometries and boundary conditions, analytical solutions to this partial differential equation (PDE) have been obtained and may be used to determine local temperatures throughout the object. When solutions do exist, in most cases they are complicated and in the form of infinite series, making computed values difficult to obtain. In most other cases of practical interest, such analytical solutions do not exist.

In many of these situations, and especially for more complicated geometries and boundary conditions, the solution can be obtained only by an approximate numerical method. This method replaces the derivatives in the above equation with approximations in the form of finite-sized differences between values at particular locations. For this reason the method is referred to as a finite difference method.

Of course, the two methods of solution—analytical and numerical—also generate different *types* of answer. In an analytical solution, a formula (usually fairly complicated) is obtained from which the temperature and its gradients can be calculated for any point in the system. In a numerical solution, on the other hand, temperatures are calculated at preselected discrete points of a grid or mesh covering the region of interest. The temperature at a point is taken as representative of a certain region which includes the point, and heat flow at a boundary is taken as the summation of the flows from the appropriate regions. The word 'point' or 'node' used here refers to a point in space for steady-state problems and to a point in space and time for unsteady transient problems.

The book attempts to provide the reader with information about all the facets of analysing and solving conductive heat transfer problems, including the computer programming aspects. The first two chapters of the book cover the fundamental ideas behind the governing equations and also revise the elements of heat transfer. These are followed by detailed explanations, in Chapters 3 and 4, of how to derive finite difference equations for a comprehensive variety of configurations. Both the Taylor series and cell control volume heat balance techniques are explained and the effect of different coordinate systems, irregular and triangular meshes, and the incorporation of boundary conditions of different types are dealt with fully. Chapters 5 and 6 show how these ideas are applied to a number of sample problems, with a coarse grid size, in order to illustrate the techniques via tubular computations and simple Fortran computer programs. There is a constant reminder that accuracy may be improved by the use of, among other things, a finer grid, so resulting in a much larger set of finite difference equations (FDEs) which gives a better simulation of the problem. Chapter 7 considers various methods of solving the resulting set of many FDEs: direct and iterative methods for steady-state and transient problems of various complexities. The merits and demerits of the methods are discussed and recommendations on efficiency (*i.e.* minimisation of computational effort required to obtain a certain accuracy) are given. Sample computations with more complicated general computer programs are now included. Miscellaneous problems are dealt with in Chapter 8 on a case study basis, amalgamating the ideas presented earlier, and these are fully documented in a consistent manner.

What special claim to originality has the present text to offer? The chief contribution of any text authors is to select, abstract and reframe that material which they find most suitable for exemplifying the principles and techniques of their field. In this undertaking the present authors have epitomised the work of several prominent writers. However, a special attribute offered by the present book is its strong practical emphasis, and recent ideas on numerical solution techniques, and their implementation via Fortran computer programs, are fully integrated in the discussion. We have tried to strike the best balance between the mathematical and physical aspects, and to produce a text which appeals to the man wishing to solve his particular practical problems.

The authors wish to acknowledge their appreciation to all those who assisted in the preparation of their textbook. We are particularly grateful: to our colleagues at Sheffield, Cranfield, Tucson and Montreal for their help, assistance, encouragement and, at times, sympathy; to the reviewers for their helpful suggestions and criticism, notably G. Hammond and J. Ward of Cranfield and M. J. Arnold and J. T. Morcomb of Tucson; to Mrs Rhoda G. Miller of Tucson for her elegant typing of most of the final manuscript; to A. E. Pratt of Milton Keynes for the preparation of many of the figures; to our wives for their patience and encouragement; and finally to Applied Science Publishers Ltd for their sympathetic cooperation and great care in the preparation of the book.

DAVID R. CROFT
DAVID G. LILLEY

Contents

CHAPTER 1

Introduction

1.1. THE BASIC IDEA

Finite difference methods offer a powerful technique for the solution of heat transfer problems. The domain in which solution is sought is replaced by a finite set of points and one tries to find approximate values for the temperature at these points. Values at these points are required to satisfy finite difference equations (FDEs) obtained either by replacing the governing partial differential equation (PDE) in terms of partial difference quotients or by direct heat flow considerations. How many points to use to achieve a certain accuracy is usually based on intuition and experience, since error estimates in practical situations are not usually available. Sometimes approximate solutions are obtained with two different sets of mesh points, one a refinement of the other, usually with the spacing halved, and the temperatures at corresponding spatial positions compared. If the change in temperature generated by the two meshes is small compared with the temperature itself then the calculations can stop, otherwise an even finer mesh should be used or both values should be used to extrapolate to a better approximation to the solution.

To clarify this basic theme, consider the problem of heat conduction in two space dimensions x and y (a 2-D problem). In steady state the temperature T satisfies a partial differential equation (PDE) of elliptic type:

$$k\left(\frac{\partial^2 T}{\partial x^2} + \frac{\partial^2 T}{\partial y^2}\right) + H = 0 \qquad (1.1.1)$$

where k is the thermal conductivity (assumed constant over the region) and H is the internal heat generation rate per unit volume. This equation can be represented by using the simplest finite difference formula for second derivatives; for the x-derivative,

$$(\Delta x)^2\left(\frac{\partial^2 T}{\partial x^2}\right)\bigg|_{(x,y)} = T(x+\Delta x, y) - 2T(x,y) + T(x-\Delta x, y) \quad (1.1.2)$$

and similarly for the y-derivative, giving the following finite difference equation (FDE), representing eqn. (1.1.1) at the point (x, y):

$$k[T(x+\Delta, y) + T(x-\Delta, y) + T(x, y+\Delta) + T(x, y-\Delta) - 4T(x,y)] + \Delta^2 H(x,y) = 0 \quad (1.1.3)$$

on taking the two spacings Δx and Δy to be equal to Δ. Of course this equation can also be derived directly from the energy balance of a small square of side Δ with centre (x, y). The energy balance states that for this 2-D steady-state problem,

Net rate at which heat is absorbed + Rate of internal heat generation = 0 (1.1.4)

and in complicated problems it is often preferable to set up an equation like eqn. (1.1.3) as a direct approximation of the physical system, via considerations like those of eqn. (1.1.4) rather than its governing differential equations, eqn. (1.1.1).

The usual boundary conditions associated with an elliptic equation are of two kinds, either:

(i) specified values of T; or
(ii) a condition on its normal gradient, usually linear and of the form

$$k\frac{\partial T}{\partial n} = -h(T - T_A) \quad (1.1.5)$$

representing linear heat transfer by convection into an ambient temperature T_A.

With elliptic equations it is necessary for boundary conditions of some type to be specified at all points of a closed curve (or surface in the case of a 3-D region) in order to define a well-posed problem in the region which it encloses. Numerical iterative solution procedures propagate the solution inwards from these boundary values.

In the case of unsteady heat flow, the difference between the heat produced and dissipated is non-zero and gives rise to a temperature change with time at any point. The associated parabolic equation is

$$\rho C_p \frac{\partial T}{\partial t} = k\left(\frac{\partial^2 T}{\partial x^2} + \frac{\partial^2 T}{\partial y^2}\right) + H \quad (1.1.6)$$

where ρ is the density, and C_p is the specific heat at constant pressure, of the material. Well-posed parabolic problems require specification of T at all points of the space region when $t = 0$, and a condition such as (i) or (ii) at all boundary points of the region when $t > 0$. For this reason they are often known as initial value problems.

Having derived the FDEs, the main problem is then that of actually computing the solution of the difference equations. For an elliptic equation the problem is that of solving a system of linear algebraic equations of the type:

$$AT + d = 0 \tag{1.1.7}$$

whereas for a parabolic equation one has to solve a system of ordinary differential equations (when some of the terms have been replaced by their finite difference approximations) of the type:

$$\frac{dT}{dt} = AT + d \tag{1.1.8}$$

where A is a square matrix, d is a column vector and T is the unknown column vector being sought. Well-posed elliptic problems generate non-singular matrices A of eqn. (1.1.7) and hence a unique solution exists. However, since there is one equation for each of (typically) several thousand mesh points, the choice of solution scheme requires great care. Because of the large number of zero elements in matrix A, iterative methods are generally used; one guesses an initial approximation to the solution and successively modifies this to a better approximation according to a given rule, until convergence is achieved to within a prescribed tolerance. By contrast, depending on what numerical procedure is used to solve a parabolic problem, represented by a set of equations like eqn. (1.1.8), it may or may not be necessary to solve a system of linear algebraic equations. With explicit methods the solution proceeds by simple replacement, but in these cases the size of the time-step is limited by stability considerations. These constraints can be partially or wholly removed by the use of implicit methods, at the expense of then having to solve systems of equations. As expected, there is often a strong relation between some of these methods for solving a transient problem and certain iterative methods for solving the associated steady-state elliptic problem.

1.2. THE PHYSICAL PROBLEM

The objective of heat conduction analysis is to predict the history of instantaneous temperature distribution and/or the fixed temperature pattern within a conducting body exposed to a different temperature environment. When this is known, other information like internal heat flow, thermal growth and thermal stresses can be derived. In practical systems there are a number of complications, including problems of realistically specifying environmental boundary conditions, the necessity to handle complicated body geometry, and the handling of a variety of material systems that differ widely from simple constant-property isotropic heat conductors.

1.2.1. Environmental Boundary Conditions

Boundary condition specification constitutes a basic input to the body response problem. One is interested in:

(i) *Interactions.* Usually the solid heat conduction problem cannot be separated from the associated problem of characterising convective and radiative surface heating conditions. Surface interactions (mass injection, combustion, etc.) couple the two problems but frequently the environment and in-depth body temperature response problems can be decoupled and treated separately. Then, as far as the body is concerned, the surface thermal boundary conditions can be prescribed (or adjusted simply according to some rule) and the problem is reduced to predicting the body temperature field subject to these constraints.

(ii) *Transients.* The general heat conduction problem is time-dependent, but if boundary conditions are time-independent, the body temperature distribution becomes steady after prolonged exposure. The exposure time required for this is short if the body conductivity k is large and vice versa. This represents the special case of steady-state heat conduction, the solution of which may be achieved by time-tracking the associated time-dependent problem, or by direct solution of the time-independent problem formulation. Conditions of steady-state temperature and heat conduction are always preceded by a starting transient t of finite duration. In some applications this damps out quickly in the heat transfer process and the body temperature field rapidly assumes a fixed steady-state pattern, whereas in other cases the early and prolonged transients are of prime importance.

(iii) *Initial and final states.* The exposure problem occurs when the body is suddenly exposed at time $t = 0$ to another environment of different temperature, and heat transfer at the surface operates to try to alter the body temperature. The body then gains or loses heat and its surface temperature tends to become equal to its surroundings, the rapidity of this change being dependent on the surface heat transfer coefficient h. For a given body conductivity k, instantaneous temperature distributions within the body depend on the relative magnitude of h/k. If this ratio is relatively large (or small), the body surface temperature tends to become equal to the ambient temperature quite rapidly (or slowly). An adiabatic (that is, lagged or insulated) boundary where no heat passes is equivalent to $h = 0$, and the normal temperature gradient $\partial T/\partial n$ is zero at such a surface.

1.2.2. Body Geometry

Some common elementary body shapes are illustrated in Fig. 1.1, and these are typical of geometric configurations that bear reasonable resemblance to actual hardware components. In fact they readily admit to exact solutions for temperature response, but numerical methods are appropriate with any more complex arrangements. They are:

- (a) infinite solid, with internal plane $x = 0$ representing a thermal contact surface;
- (b) semi-infinite solid, with environment exposed face at $x = 0$;
- (c) insulated plate, with finite width exposed at one side and insulated at the other;
- (d) composite plate (or semi-infinite solid if δ_2 is large) with finite thermal contact resistance between the layers 1 and 2 with a sudden temperature drop;
- (e) cylindrical or spherical cavities, with internal environment;
- (f) cylinder or sphere, with external environment;
- (g) cylindrical or spherical shell, with inner or outer surface thermally insulated;
- (h) elliptical solid, with external environment;
- (i)–(k) extended surfaces, which are finned or spined bodies whose purpose is to increase heat exchange between an otherwise simple surface and its hostile environment.

In each of examples (a)–(d), a typical temperature plot is superimposed.

1.2.3. Material Systems

Practical materials and structures are more complicated than their idealised mathematical models, and this often presents the most difficult aspect of thermal-structural analysis. The nature and diversity of typical materials and structures used in practice is vast, and for further information the reader is recommended to refer to McAdams (1954).

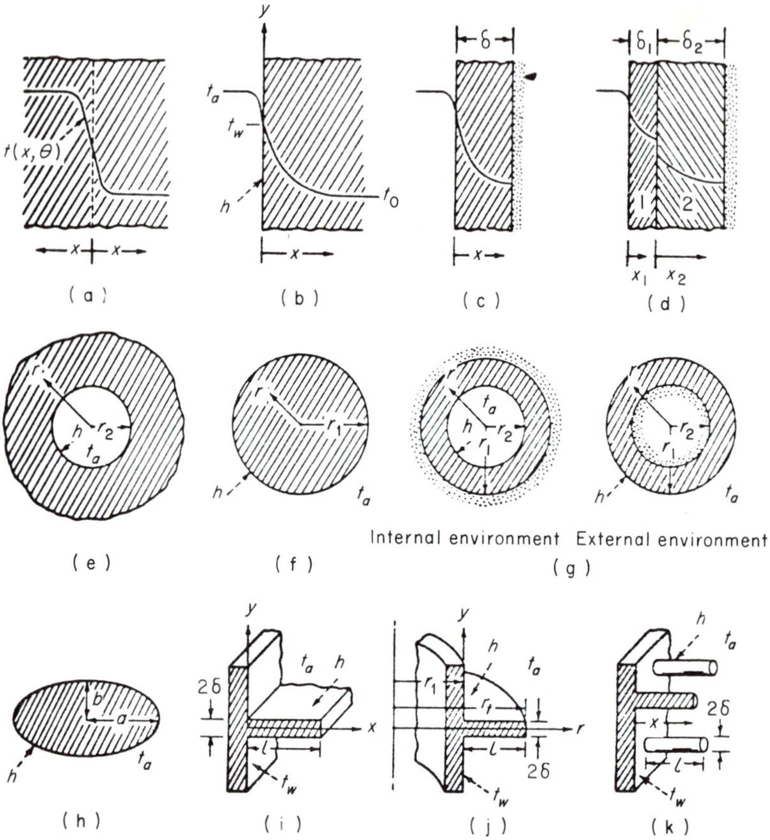

FIG. 1.1. Elementary body shapes: (a) infinite solid, (b) semi-infinite solid, (c) plate, (d) composite plate, (e) cylindrical and spherical cavity, (f) cylinder and sphere, (g) cylindrical and spherical shell, (h) ellipse and ellipsoid, (i) straight fin, (j) annular fin, (k) spine. (Based on Schneider, 1973, Fig. 2, p. 3–5. © W. M. Rohsenow and J. P. Hartnett 1973. Used with permission of McGraw-Hill Book Company.)

1.3. THE MATHEMATICAL PROBLEM

Typical of many problems in mathematical physics, heat transfer problems can be formulated as second-order partial differential equations (PDEs) of the form

$$\operatorname{div}\,(k \text{ grad } T) + H = \rho C_{\mathrm{p}} \frac{\partial T}{\partial t} \qquad (1.3.1)$$

which must be solved, in some spatial region R within a boundary B, subject to known boundary conditions of the form:

(i) T is known on B;

(ii) $\beta \dfrac{\partial T}{\partial n} + \gamma \dfrac{\partial T}{\partial s}$ is known on B; or (1.3.2)

(iii) $\alpha T + \beta \dfrac{\partial T}{\partial n} + \gamma \dfrac{\partial T}{\partial s}$ is known on B.

These three types of boundary conditions are called the Dirichlet, the Neumann and the third boundary condition, respectively. When different conditions of eqn. (1.3.2) hold on different parts of B, the boundary conditions are said to be mixed. Condition (i) applies when the boundary B is maintained at a known temperature, whereas the familiar form

$$\frac{\partial T}{\partial n} = -\frac{h}{k}(T - T_A) \tag{1.3.3}$$

representing heat transfer by convection from the surface into an ambient temperature T_A, is an example of condition (ii). The example quoted is the general heat conduction equation where k = conductivity, T = temperature, H = internal heat generation per unit volume per unit time, ρ = density, C_p = specific heat at constant pressure, t = time, α, β, γ are scalars, n = normal direction to surface, s = tangential direction on surface, and h = heat transfer coefficient.

The most general PDE for heat conduction is eqn. (1.3.1) with k, H, ρ and C_p being spatial, time and temperature dependent, that is:

$$k = k(\mathbf{x}, t, T); \quad H = H(\mathbf{x}, t, T); \quad \rho = \rho(\mathbf{x}, t, T); \quad C_p = C_p(\mathbf{x}, t, T) \tag{1.3.4}$$

where $\mathbf{x}$ represents some point in three-dimensional (3-D) space and k may also depend on direction (non-isotropic medium). Equation (1.3.1) includes the majority of special cases of practical interest. For example, if the thermal conductivity k of the system is constant and isotropic (*i.e.* has the same value in all directions), then the temperature $T = T(\mathbf{x}, t)$ must satisfy the equation

$$\nabla^2 T + \frac{H}{k} = \frac{1}{\alpha}\frac{\partial T}{\partial t} \tag{1.3.5}$$

where $\nabla^2 \equiv$ div grad (Laplacian operator), and $\alpha = k/\rho C_p$ is the thermal diffusivity, a property of the conducting material. Further, if the system does not contain heat sources or sinks ($H = 0$), $T = T(\mathbf{x}, t)$ must satisfy the diffusion equation

$$\nabla^2 T = \frac{1}{\alpha}\frac{\partial T}{\partial t} \tag{1.3.6}$$

or, if sources or sinks are present but the temperature is steady ($\partial T/\partial t = 0$), then $T = T(\mathbf{x})$ must satisfy the Poisson equation

$$\nabla^2 T + \frac{H}{k} = 0 \tag{1.3.7}$$

Finally, for steady conduction ($\partial T/\partial t = 0$) in systems free of sources and sinks ($H = 0$), $T = T(\mathbf{x})$ must satisfy the Laplace equation

$$\nabla^2 T = 0 \tag{1.3.8}$$

Equations like eqn. (1.3.1) arise in the study of many engineering phenomena of practical importance. Similar PDEs occur, for example, in the following fields:

(a) Electrical fields in dielectrics.
(b) Electrical fields in conducting media.
(c) Magnetic fields.
(d) Motion of non-viscous and viscous fluids.
(e) Diffusion.
(f) Structures in equilibrium.

Indeed the analogy between these fields is often exploited in the simulation and solution of a problem from a different field. For example, an analogue experiment to simulate a heat transfer problem might take advantage of the mathematical analogy between heat flow and the familiar electrical potential-field phenomena. Further details may be found in Arpaci (1966), Gebhart (1971) and Schneider (1955, 1973).

All the above PDEs with $H = 0$ possess the trivial solution

$$T \equiv 0 \tag{1.3.9}$$

which will rarely constitute a solution of the physical problem since usually $T \neq 0$ on the boundary. Solutions must be found which satisfy the additional constraints, called boundary conditions, which are an inherent part of even the general solution to the problem.

Problems represented by the basic equation, eqn. (1.3.1), and its degenerate forms, are generally classified as:

(i) elliptic boundary value problems, if the equilibrium steady-state ($\partial T/\partial t = 0$) form of the equation is being considered: or

(ii) parabolic initial boundary value problems, if the unsteady-state ($\partial T/\partial t \neq 0$) form of the equation is being considered.

The difference between these classes concerns the analytic character of their solutions and the types of boundary conditions necessary to determine these solutions. The significance of this difference may be outlined as follows:

(a) The usual boundary conditions associated with an elliptic steady-state problem are of two kinds. Often one has conditions of type (i) or (ii) of eqn. (1.3.2) at all points of the closed boundary B (or a combination of these on parts of B), the specification being of type (i) on at least part of B. These conditions generally suffice to define a well-posed problem in the region R which the boundary B encloses.

(b) In the case of a parabolic unsteady-state problem, the temperature T at any point in the region R may change with position $\mathbf{x}$ as well as with time t. The boundary conditions required for such a problem to be well posed are that T be specified at all points of the spatial region R when $t = 0$, and that a condition such as (i) or (ii) of eqn. (1.3.2) be given at all points on the boundary B of the region R when $t > 0$.

1.4. METHODS OF SOLUTION

In order to evaluate these temperature fields in conducting bodies, there are basically four available methods.

1.4.1. Analytical Methods

These involve the derivation of the mathematical solution for the temperature as a function of space or space–time coordinates. The solution must satisfy the basic governing partial differential equation, together with certain initial and boundary conditions appropriate to the particular problem. Usually the actual problem must be simplified in order for this approach to succeed. Thus the solution so found is by no means exact, as far as the physical problem is concerned. But such solutions are often preferable, inasmuch as they allow easy interpretation of the effect of slight problem changes. Basic texts include Carslaw & Jaeger (1959) and Crank (1956).

1.4.2. Graphical Methods

Graphical methods are based on the properties of the characteristic field equations and numerical principles, and have the singular advantage of giving a rapid first approximation to the required solution.

1.4.3. Numerical Methods

Numerical methods are usually based on finite difference approximations and hence derive an approximate computational solution, but one whose accuracy can usually be increased to any desired degree. The techniques used are fairly straightforward and generally applicable, and yield answers to problems which are often awkward and sometimes entirely intractable by formal analytical means. They suffer, of course, from the lack of parameterisation that formal solutions possess.

1.4.4. Experimental Methods

These are normally reserved for problems which cannot be handled successfully or conveniently by the previous three methods. Use may be made of actual models on which measurements are made; alternatively, analogue experiments may be employed, based on the mathematical analogy between heat flow and other familiar potential-field phenomena.

It is useful to be familiar with the power and limitations of all these techniques, so that with each new problem one can attempt the most realistic method of solution with regard to conservation of time, labour, expense, talent and accuracy. Despite this, our text is devoted solely to finite difference numerical solution methods. The methods are basically simple but offer a most powerful tool to the engineering designer or researcher faced with heat transfer problems of a difficult, or more often impossible, analytical nature. Nor can these tasks be relegated from the engineer to the computer programmer—normally the physical problem and the governing equations require, in any except the simplest problems, the engineer's experience and intuition—so there is a real need for the engineer to understand these techniques.

1.5. DISCRETISATION

When a finite difference technique is to be used to solve a PDE (plus associated boundary and possibly initial conditions), a network of grid points is first established throughout the region R of interest. These grid points are usually taken as points of intersection of a grid covering the region. Often the words 'mesh points', 'pivotal points', 'nodal points' and 'lattice points' are used instead of 'grid points' and the words 'mesh' and 'lattice' instead of 'grid'. Figure 1.2 shows how the basic approximation of replacing the continuous region R by a set of discrete points P_{ij} may be accomplished in the case of a two-dimensional (2-D) region. In the case of a

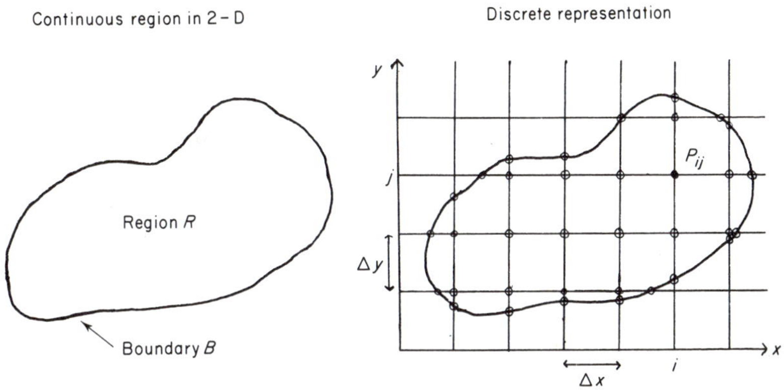

FIG. 1.2. Discrete approximation of a continuous 2-D region.

transient problem a third coordinate t is taken outwards from the page and the 2-D grid is repeated at intervals of Δt outwards, making a 3-D grid system. To be specific, suppose there are two distance coordinates x and y and time t as independent variables and that the respective grid spacings are Δx, Δy and Δt. Subscripts i, j and m may then be used to denote that point with space coordinates $i\Delta x$, $j\Delta y$, $m\Delta t$, also called the grid point $(i,j,m) = P_{i,j,m}$. Instead of developing a solution which is valid everywhere in R (for example, like an analytic solution), only approximations are obtained at the isolated points $P_{i,j,m}$. Let the exact solution to the PDE be $T = T(x, y, t)$ and its approximation, to be determined at each grid point by the method of finite differences, be $U = U_{i,j,m}$. Then the exact solution $T = T(i\Delta x, j\Delta y, m\Delta t) = T_{i,j,m}$ at a particular grid point (i, j, m) will in general differ from $U_{i,j,m}$.

To obtain the approximate solution $U_{i,j,m}$ at a grid point (i,j,m), the original PDE with associated boundary conditions is approximated at all points $P_{i,j,m}$ by suitable finite difference expressions involving Δx, Δy, Δt and approximate solutions $U_{i,j,m}$ at several neighbouring grid points. This procedure leads to a set of linear algebraic equations called finite difference equations (FDEs) in the unknowns $U_{i,j,m}$, one such equation for each point of the grid system, which represents the single PDE. This set of FDEs may then be solved to a certain accuracy. Then intermediate values, integrals, derivatives, heat flows, etc., may be obtained from the discrete solution at the points $P_{i,j,m}$ by numerical interpolation techniques. By making the grid spacing sufficiently small it is expected that the approximate solution will be sufficiently accurate. This deceptively simple technique has many pitfalls—there are many serious and difficult mathematical questions of adequacy, accuracy, convergence and stability, which will be dealt with later. Further information may also be found in texts on the numerical solution of partial differential equations, for example Ames (1969).

In the above paragraph the word 'suitable' appeared in connection with the finite difference representation of the problem at a point $P_{i,j,m}$. What constitutes 'suitable' is an important discretisation problem, and both mathematical and physical techniques for discretisation abound in the literature. In the mathematical approach the continuous formulation of the problem, eqns. (1.3.1) and (1.3.2), is transformed into a discrete formulation by replacing derivatives and other functions occurring by finite difference approximations. The technique is simple, straightforward and flexible—one can assign an order of accuracy to each replacement, for example—and finite difference methods (via Taylor series or finite difference operators), variational methods and the method of lines are used to set up the FDEs. In complicated problems the specialist may prefer to set up equations at grid points by the direct approximation to the laws governing the physical system, rather than to the PDE itself. This resistance network or cell control volume approach can be of great value in difficult cases which are immediately simplified by the examination of heat flows in a control volume, especially at points on or near, for example, irregular boundaries with convective heat loss. However, in some respects, this somewhat physical and intuitive approach can sometimes produce approximations of low accuracy and poor consistency with the problem. Though the order of accuracy cannot be easily determined, results are often the same as with the finite difference method. In the following pages, either or both of these techniques will be used.

BIBLIOGRAPHY

AMES, W. F. (1969). *Numerical Methods for Partial Differential Equations*. Nelson, London.

ARPACI, V. S. (1966). *Conduction Heat Transfer*. Addison-Wesley, Reading, Mass.

BAYLEY, F. J., OWEN, J. M. & TURNER, A. B. (1972). *Heat Transfer*. Nelson, London.

CARSLAW, H. S. & JAEGER, J. C. (1959). *Conduction of Heat in Solids*. Clarendon Press, Oxford.

CRANK, J. (1956). *Mathematics of Diffusion*. Clarendon Press, Oxford.

GEBHART, B. (1971). *Heat Transfer*, 2nd ed. McGraw-Hill, New York.

HOLLAND, F. A., MOORES, R. M., WATSON, F. A. & WILKINSON, J. K. (1970). *Heat Transfer*. Heinemann, London.

MCADAMS, W. H. (1954). *Heat Transmission*, 3rd ed. McGraw-Hill, New York.

RAZELOS, P. (1973). Methods of obtaining approximate solutions. Section 4, pp. 4–1 to 4–78, in *Handbook of Heat Transfer*, ed. W. M. Rohsenow and J. P. Hartnett. McGraw-Hill, New York.

ROHSENOW, W. M. & HARTNETT, J. P. (eds.) (1973). *Handbook of Heat Transfer*. McGraw-Hill, New York.

SCHNEIDER, P. J. (1955). *Conduction Heat Transfer*. Addison-Wesley, Reading, Mass.

SCHNEIDER, P. J. (1973). Conduction. Section 3, pp. 3–1 to 3–134, in *Handbook of Heat Transfer*, ed. W. M. Rohsenow and J. P. Hartnett. McGraw-Hill, New York.

SCHUH, H. (1965). *Heat Transfer in Structures*. Pergamon Press, Oxford.

STEPHENSON, G. (1970). *An Introduction to Partial Differential Equations for Science Students*, 2nd ed. Longman, London.

SUCEC, J. (1975). *Heat Transfer*. Simon & Schuster, New York.

WALSH, J. (1966). Elliptic and parabolic equations. Chapter 5, pp. 99–116, in *Numerical Analysis: An Introduction*, ed. J. Walsh. Academic Press, London.

CHAPTER 2

Fundamentals

2.1. BASIC EQUATIONS OF HEAT TRANSFER

In most industrial processes, heat is usually transferred from one part to another by a combination of modes: conduction, convection and radiation. It depends upon the level of temperature and types of substances involved as to which particular mode is important. For ease of explanation, it is usual to consider the modes of heat transfer separately.

2.1.1. Conduction

Heat is transferred by conduction through a solid body or a fluid at rest. In the case of a solid, energy is transferred by the movement of electrons within the lattice of fixed molecules, whereas for a fluid there will be direct molecular interaction as the molecules are free to move.

Newton, and later Fourier, evolved the basic Fourier law of heat conduction from experimental observations on steady-state systems. They found that

Heat flow $\propto$ (temperature difference)
$\propto$ (area normal to direction of heat flow)

giving the equation

$$Q_x = -kA\frac{\partial T}{\partial x} \tag{2.1.1}$$

where Q_x is the rate of heat flow in the x-direction (W = J/s), A is the area normal to the x-direction (m^2), $\partial T/\partial x$ is the temperature gradient in the x-direction (K/m), and k is the constant of proportionality, known as the *thermal conductivity* of the particular substance involved (W/m K). (The negative sign is required to make the equation consistent, *i.e.* that heat will flow into a region of lower temperature, giving a positive value for Q_x when $\partial T/\partial x$ is negative.)

For a parallel-sided wall of thickness (Δx) of thermal conductivity k, subjected to temperatures T_1 and T_2 and assuming heat flows only in the x-direction, the equation becomes

$$Q_x = kA\frac{(T_1 - T_2)}{\Delta x} \tag{2.1.2}$$

and is illustrated in Fig. 2.1.

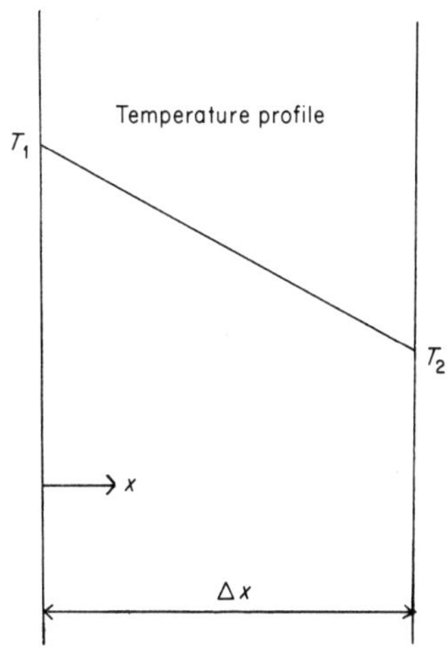

FIG. 2.1. 1-D conduction through a plane wall.

Equation (2.1.1) is often expressed in the alternative form

$$Q''_x = -k\frac{\partial T}{\partial x} \tag{2.1.3}$$

where $Q''_x = Q_x/A$ is the heat flux (W/m^2).

In general, Fourier's law, eqn. (2.1.1), has certain limitations and does not hold if:

(i) the thermal conductivity is not constant—then it holds only roughly over small intervals Δx;

(ii) the thermal conductivity is direction-dependent—then the k appropriate to the direction being considered must be used;

(iii) the heat flows other than in the x-direction, in the case of the form of the law quoted above; or

(iv) a steady state does not exist, *i.e.* if the temperature at a particular location changes with time.

Several points may be noted about the value of the thermal conductivity k:

(i) metals have a high value of k representing good conductivity;

(ii) liquids and gases have low values of k representing poor conductors;

(iii) k for an alloy cannot be determined from the simple weighted average of the conductivity values of its constituents;

(iv) for pure metals k decreases with temperature;

(v) many insulating compounds rely on their porosity for their insulation effect and there is an optimum packing for maximum insulation (minimal overall k)—if air spaces are too large then convection heat transfer occurs, but if they are too small then conduction through the solid connections predominates; and

(vi) for materials such as graphite, k is extremely directionally sensitive.

Typical values of k for a variety of materials are illustrated in Table 2.1.

TABLE 2.1
TYPICAL THERMAL CONDUCTIVITIES FOR A VARIETY OF MATERIALS AT 288 K

Substance	k (W/mK)
Silver	415·0
Copper	380·0
Mild steel	45·0
Stainless steel	19·0
Wood	0·17
Asbestos	0·17
Water	0·60
Air	0·026

In the general analysis the thermal conductivity k at a point $\mathbf{x} = (x, y, z)$ will be considered to depend upon the orientation of the direction for which Fourier's law, eqn. (2.1.1), is written. The simplest situation is that in which k, though it may vary with position $\mathbf{x}$ and temperature T, is the same in all directions for each point being considered.

This is the case in gases, and in most liquids and solids. Then the equation

$$Q''_{x_j} = -k\frac{\partial T}{\partial x_j} \tag{2.1.4}$$

where x_j is either x, y or z, holds as in eqn. (2.1.1). In this case the heat-flux vector Q'' is given in terms of the gradient of the temperature field ∇T as

$$Q'' = -k\nabla T \tag{2.1.5}$$

However, the more general non-isotropic formulation of conduction postulates that any component of the heat-flux vector depends on the temperature gradients in each of the three coordinate directions so that, for the x-direction for example,

$$Q''_x = -k_{11}\frac{\partial T}{\partial x} - k_{12}\frac{\partial T}{\partial y} - k_{13}\frac{\partial T}{\partial z} \tag{2.1.6}$$

Three conductivity components arise in each coordinate direction and the thermal conductivity becomes the following tensoral quantity:

$$k_{ij} = \begin{bmatrix} k_{11} & k_{12} & k_{13} \\ k_{21} & k_{22} & k_{23} \\ k_{31} & k_{32} & k_{33} \end{bmatrix} \tag{2.1.7}$$

Thus eqn. (2.1.6) may be alternatively written as

$$Q''_{x_i} = -k_{i1}\frac{\partial T}{\partial x} - k_{i2}\frac{\partial T}{\partial y} - k_{i3}\frac{\partial T}{\partial z} \tag{2.1.8}$$

or simply as

$$Q''_{x_i} = -k_{ij}\frac{\partial T}{\partial x_j} \tag{2.1.9}$$

where the summation convention (summing on repeated suffixes—j in this case) has been used.

The general non-isotropic case includes the isotropic case if $k_{ij} = 0$ $(i \neq j)$, and $k_{ii} = k$; that is, all non-diagonal terms in the tensoral formulation above are zero and the remaining non-zero diagonal ones are all equal. Another subcase often arises in ordered materials such as wood, fibrous materials and crystalline substances (for example graphite) when the conductivity is directionally dependent. If coordinate axes are aligned with the principal axes of conductivity of the material, then $k_{ij} = 0$ $(i \neq j)$, and $k_{ii} = k_i$; that is, all non-diagonal terms are zero and the diagonal terms are different, representing different conductivities in different directions.

The heat-flux components are then

$$Q''_x = -k_x \frac{\partial T}{\partial x}; \quad Q''_y = -k_y \frac{\partial T}{\partial y}; \quad Q''_z = -k_z \frac{\partial T}{\partial z} \qquad (2.1.10)$$

Not all problems may be reduced in this manner. For example, if boundary conditions have to be applied on surfaces which are not normal to the coordinate axes, there are difficulties in applying the boundary conditions. When transient conditions prevail, the temperature of the substances at a given point will be changing with time: the magnitude of the variation will be a function of the specific heat C_p of the material. In mathematical terms:

$$\text{Net rate of heat flow } Q = mC_p \frac{\partial T}{\partial t} \quad \text{(W)} \qquad (2.1.11)$$

where m is the mass of the substance (kg), C_p is the specific heat of the substance (J/kg K), and $\partial T/\partial t$ is the rate of temperature change (K/s). In turn, the net rate of heat flow to a point will be the difference between the heat conducted to, and the heat conducted away from, the point. To summarise, the specific heat is a measure of capacity to store heat whereas thermal conductivity indicates the ability of a material to transmit heat. In a transient system the rate of response of a system to a temperature change will be a function of both these properties and the mass of the material. The term 'thermal diffusivity $\propto$' has been evolved to partially quantify the rate of response:

$$\text{Thermal diffusivity } \propto = \frac{k}{\rho \cdot C_p} \quad (\text{m}^2/\text{s}) \qquad (2.1.12)$$

where ρ is the density of the material (kg/m^3).

A dimensionless group, called a Fourier number (Fo), has been defined to assist in the representation of transient heat flow calculations:

$$\text{Fo} = \frac{\propto \cdot \Delta t}{(\Delta x)^2} = \frac{k\,\Delta t}{\rho C_p (\Delta x)^2} = \frac{\text{Rate of conduction of heat}}{\text{Rate of storage of heat}} \qquad (2.1.13)$$

where Δt is the increment of time being considered (s), and (Δx) is the thickness of the substances (or segment of substance) being considered (m). The Fourier number (Fo) gives an indication of the speed at which a body will respond to a temperature change. Low values of Fo imply that a long period of time is required to heat or cool the body, and vice versa.

2.1.2. Convection

Convective heat transfer takes place when whole groups of molecules move from one place to another of different temperature. Convection is thus limited to liquids and gases and is normally concerned with heat exchange between fluids and solid boundaries.

Where the motion of the fluid is induced by buoyancy forces, *e.g.* a hot plate in air, when density changes occur due to temperature increases, then the mode of heat transfer is called 'free' convection. The rates of heat transfer by this mode are dependent on the particular properties of the fluid. When the motion of the fluid is externally induced, by a fan for example, then although a small amount of free convection may be present, the mode is said to be 'forced' convection.

It was Newton again who observed that heat transfer by convection was a function of the temperature difference between the solid and the fluid T_s and T_f and the area of contact A:

$$\text{Heat flow } Q \propto (T_s - T_f)$$
$$\propto A$$

In fact

$$Q = hA(T_s - T_f) \quad \text{(W)} \qquad (2.1.14)$$

where the constant of proportionality h is called the heat transfer coefficient (W/m^2K). The determination of h for a variety of fluids and conditions has been the object of experimental work for many years, leading to a whole variety of relationships between h and the flow parameters (velocity of flow, condition and size of surface of contact, fluid properties, etc.). In many cases the results have been correlated by dimensional analysis, generating expressions of the type

$$\text{Nu} = \text{const } (\text{Re})^n(\text{Pr})^m \qquad (2.1.15)$$

where

$$\text{Nu} = \text{Nusselt number} = \frac{h \times \text{A characteristic dimension}}{\text{Thermal conductivity of fluid}};$$

$$\text{Re} = \text{Reynolds number} = \frac{\text{Velocity} \times \text{Dimension}}{\text{Kinematic viscosity}}; \text{ and}$$

$$\text{Pr} = \text{Prandtl number} = \frac{\text{Specific heat} \times \text{Dynamic viscosity}}{\text{Thermal conductivity}}$$

Where convection occurs with fluids of low boiling point, then boiling or condensation may take place. Heat transfer involving change of phase usually generates very high heat transfer coefficients due to high latent heats involved. The complex motion, in boiling particularly, has so far defied analysis and predictions of heat transfer rates involving change of phase are somewhat empirical. However, to give some idea of the relative rates of heat transfer by convection in its various modes, a list of typical coefficients is given in Table 2.2. At the end of the table, the coefficients for dropwise condensation, where the film of condensation breaks up and falls off the condensing surface, indicate that the resistance to heat flow is virtually nil.

TABLE 2.2

TYPICAL HEAT TRANSFER COEFFICIENTS FOR A VARIETY OF SITUATIONS

		$h\ (W/m^2K)$
Free convection:	gases	1–100
	(air	10–15)
	liquids	50–1 000
	(water	500–700)
Forced convection:	gases	10–500
	(air	20–300)
	liquids	100–10 000
	(water	up to 1 000)
Boiling:		500–20 000
Condensation:		500–30 000 (filmwise)
		20 000–500 000 (dropwise)

In connection with the relative magnitude of surface conductance to internal conduction, the Biot number (Bi) is a dimensionless group defined by

$$\mathrm{Bi} = \frac{h\,\Delta x}{k} = \frac{\text{Surface conductance}}{\text{Thermal conductivity of solid}} \tag{2.1.16}$$

2.1.3. Radiation

Radiation, unlike conduction and convection, does not require a solid or fluid for its propagation; radiant energy is transmitted by electromagnetic waves. The characteristics of those waves which transmit thermal energy as opposed to those concerned in radio, X-rays or gamma-rays are simply in the frequency and wavelength. Thermal radiation can be propagated, reflected, refracted, etc., in exactly the same way as visible electromagnetic radiation (light) and will travel through space without the assistance of other material (see Gray & Müller, 1974). The actual amount of energy transmitted by this mode varies with the temperature of the emitting body, the wavelength of the emitted radiation, and the nature and area of the surface of the emitter.

It has been established that, given perfect conditions, the rate at which energy is radiated from a body is proportional to the fourth power of its absolute temperature, *i.e.*

$$Q \propto T^4$$

$$Q = \sigma T^4 \quad (\mathrm{W/m^2}) \tag{2.1.17}$$

where σ, the Stefan–Boltzmann constant, $= 5{\cdot}67 \times 10^{-8}\,\mathrm{W/m^2K^4}$. If a body did radiate this quantity of heat (no body in practice actually does), it would be called a blackbody. In fact all materials give a fraction of blackbody radiation, *i.e.*

$$Q = E\sigma T^4 \quad (\mathrm{W/m^2}) \tag{2.1.18}$$

where the fraction $E(<1)$ is termed the emissivity. This emissivity is largely a function of surface finish: for copper polished $E = 0{\cdot}02$, for copper oxidised $E = 0{\cdot}7$, and for copper treated with a matt black finish $E = 0{\cdot}95$.

In practice we are normally concerned with radiation between two systems say at temperatures T_1 and T_2, where T_1 is greater than T_2. The body at T_2 will emit some radiation, but it will receive more radiation from T_1 and hence will increase in temperature. The net heat received by body 2

$$= \text{Heat received from body 1} - \text{Heat transmitted by body 2}$$

$$= F_{12}EA\sigma(T_1^4 - T_2^4) \quad (\mathrm{W}) \tag{2.1.19}$$

where A is the receiving area of body 2, and F_{12} is a geometric factor between bodies 1 and 2.

The primary concern here is to remind the reader of the basic expressions involved in heat transfer. The conduction formulae will clearly be relevant in the non-flow conditions pertaining inside solid objects, and the convective and radiative flows will occur at the boundaries.

2.2. GENERAL EQUATION OF HEAT FLOW

The partial differential equation of conduction is derived by considering the energy balance (heat flows) on a differential element of volume ($\mathrm{d}x\,\mathrm{d}y\,\mathrm{d}z$) of a material subject to a conduction process and shown in Fig. 2.2. The net energy gain in time δt across the six faces by conduction is added to the energy generation in time δt and the sum is set equal to the change in time δt of energy stored in the mass inside the volume element, as indicated by its temperature rise. To do this, the appropriate form of Fourier's law of conduction from Section 2.1 is used along with eqn. (2.1.11). Assuming there is no movement of mass across the faces of the element (that is, velocity is zero), these terms are the only ones contributing to the energy balance. The discussion follows Gebhart (1971).

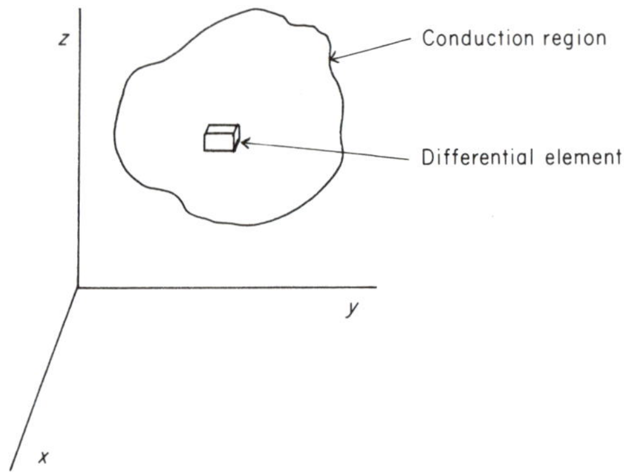

FIG. 2.2. Differential element for conduction analysis.

The midpoint of the differential element is denoted as x, y, z, and at time t the temperature and specific heat at this point are T and C_p. The density ρ is considered independent of time. The energy generation is assumed due to a source distributed throughout the material, and its strength, H, is the rate of energy generation at x, y, z at time t per unit volume. This rate may also depend upon local temperature. Therefore, in general, $H = H(x, y, z, T, t)$.

The general equation is derived for four levels of complexity, dependent upon the degree of variability allowed for the thermal conductivity k. In each case the density ρ and specific heat C_p may be functions of position, temperature or time, the relations

$$\rho = \rho(x, y, z, T, t); \quad C_p = C_p(x, y, z, T, t)$$

being appropriate. The following forms for k are considered (see Section 2.1):

(i) *Constant and isotropic.* k = constant, in which case Fourier's law of heat conduction may be taken as

$$Q_x = -kA\frac{\partial T}{\partial x}; \quad Q_y = -kA\frac{\partial T}{\partial y}; \quad Q_z = -kA\frac{\partial T}{\partial z}$$

(ii) *Variable.* k dependent on position, temperature or time, taking here $k = k(T)$ to simplify expressions and then

$$Q_x = -k(T)A\frac{\partial T}{\partial x}; \quad Q_y = -k(T)A\frac{\partial T}{\partial y};$$

$$Q_z = -k(T)A\frac{\partial T}{\partial z}$$

(iii) *Orthotropic.* $k = k_x, k_y$ or k_z dependent upon which direction (x, y or z) heat is being conducted and then

$$Q_x = -k_xA\frac{\partial T}{\partial x}; \quad Q_y = -k_yA\frac{\partial T}{\partial y}; \quad Q_z = -k_zA\frac{\partial T}{\partial z}$$

allowing for the k_x, k_y and k_z to vary as in (ii) also.

(iv) *Nonisotropic.* k has nine components k_{ij} ($1 \leq i, j \leq 3$) related to the Q's and x_1, x_2, x_3 ($\equiv x$, y, z) by the expressions

$$Q_x = -k_{11}A\frac{\partial T}{\partial x} - k_{12}A\frac{\partial T}{\partial y} - k_{13}A\frac{\partial T}{\partial z}$$

$$Q_y = -k_{21}A\frac{\partial T}{\partial x} - k_{22}A\frac{\partial T}{\partial y} - k_{23}A\frac{\partial T}{\partial z}$$

$$Q_z = -k_{31}A\frac{\partial T}{\partial x} - k_{32}A\frac{\partial T}{\partial y} - k_{33}A\frac{\partial T}{\partial z}$$

or, written simply, as

$$Q_{x_i} = -k_{i1}A\frac{\partial T}{\partial x} - k_{i2}A\frac{\partial T}{\partial y} - k_{i3}A\frac{\partial T}{\partial z}$$

or

$$Q_{x_i} = -k_{ij}A\frac{\partial T}{\partial x_j}$$

where the summation convention, summing the terms with a repeated suffix, in this case $j = 1$, 2 and 3, has been assumed.

In each case ((i), (ii), (iii) or (iv)) the derivation of the general equation is very similar. Here all four cases are derived simultaneously and one merely reads the appropriate line ((i), (ii), (iii) or (iv) according to the form desired by the reader) of each subsequent equation (eqns. (2.2.1)–(2.2.4)).

The general equation for the 3-D flow of heat by conduction may be derived by considering the heat flow firstly in the x-direction, as shown in

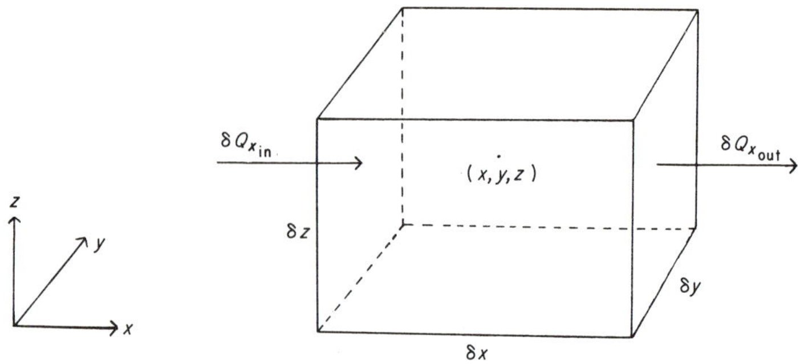

FIG. 2.3. Heat conduction through a differential element.

Fig. 2.3. The heat flows into and out of the element in time δt are given, by application of the appropriate form of Fourier's law, as

(i) $$\delta Q_{x_{in}} = -k(\delta y \delta z)\left(\frac{\partial T}{\partial x}\right)_{x-\delta x/2} \cdot \delta t \qquad (2.2.1)$$

(ii) $$= -(\delta y \delta z)\left(k(T)\frac{\partial T}{\partial x}\right)_{x-\delta x/2} \cdot \delta t$$

(iii) $$= -(\delta y \delta z)\left(k_x \frac{\partial T}{\partial x}\right)_{x-\delta x/2} \cdot \delta t$$

(iv) $$= -(\delta y \delta z)\left(k_{1j} \frac{\partial T}{\partial x_j}\right)_{x-\delta x/2} \cdot \delta t$$

and similar expressions with $x + \delta x/2$ replacing $x - \delta x/2$ for the evaluation of $\delta Q_{x_{out}}$.

A simple application of Taylor series reveals that

$$\left(\quad\right)_{x+\delta x/2} \approx \left(\quad\right)_{x-\delta x/2} + \delta x \cdot \frac{\partial}{\partial x}\left(\quad\right)$$

where the right-hand term may be evaluated, for the purpose of the discussion here, at the centre point x. For example

$$\left(\frac{\partial T}{\partial x}\right)_{x+\delta x/2} \approx \left(\frac{\partial T}{\partial x}\right)_{x-\delta x/2} + \delta x \cdot \frac{\partial}{\partial x}\left(\frac{\partial T}{\partial x}\right)$$

and this may be used in connection with the (i) formulae. Hence the net amount of heat absorbed by the element in time δt because of x-direction heat flow is

$$\delta Q_{x_{net}} = \delta Q_{x_{in}} - \delta Q_{x_{out}}$$

so that

$$\text{(i)} \qquad \delta Q_{x_{\text{net}}} = k(\delta x \delta y \delta z)\frac{\partial^2 T}{\partial x^2} \cdot \delta t \qquad (2.2.2)$$

$$\text{(ii)} \qquad = (\delta x \delta y \delta z)\frac{\partial}{\partial x}\left(k(T)\frac{\partial T}{\partial x}\right) \cdot \delta t$$

$$\text{(iii)} \qquad = (\delta x \delta y \delta z)\frac{\partial}{\partial x}\left(k_x\frac{\partial T}{\partial x}\right) \cdot \delta t$$

$$\text{(iv)} \qquad = (\delta x \delta y \delta z)\frac{\partial}{\partial x}\left(k_{1j}\frac{\partial T}{\partial x_j}\right) \cdot \delta t$$

Similar analyses in the other two directions give corresponding expressions so that the total heat absorbed by the element in time δt because of heat flow is

$$\text{(i)} \quad \delta Q_{\text{flow}} = k(\delta x \delta y \delta z)\left(\frac{\partial^2 T}{\partial x^2} + \frac{\partial^2 T}{\partial y^2} + \frac{\partial^2 T}{\partial z^2}\right) \cdot \delta t \qquad (2.2.3)$$

$$\text{(ii)} \qquad = (\delta x \delta y \delta z)\left[\frac{\partial}{\partial x}\left(k(T)\frac{\partial T}{\partial x}\right) + \frac{\partial}{\partial y}\left(k(T)\frac{\partial T}{\partial y}\right) + \frac{\partial}{\partial z}\left(k(T)\frac{\partial T}{\partial z}\right)\right] \cdot \delta t$$

$$\text{(iii)} \qquad = (\delta x \delta y \delta z)\left[\frac{\partial}{\partial x}\left(k_x\frac{\partial T}{\partial x}\right) + \frac{\partial}{\partial y}\left(k_y\frac{\partial T}{\partial y}\right) + \frac{\partial}{\partial z}\left(k_z\frac{\partial T}{\partial z}\right)\right] \cdot \delta t$$

$$\text{(iv)} \qquad = (\delta x \delta y \delta z)\left[\frac{\partial}{\partial x}\left(k_{1j}\frac{\partial T}{\partial x_j}\right) + \frac{\partial}{\partial y}\left(k_{2j}\frac{\partial T}{\partial x_j}\right) + \frac{\partial}{\partial z}\left(k_{3j}\frac{\partial T}{\partial x_j}\right)\right] \cdot \delta t$$

[The last one of these may also be written

$$\delta Q_{\text{flow}} = (\delta x \delta y \delta z)\left[\frac{\partial}{\partial x_i}\left(k_{ij}\frac{\partial T}{\partial x_j}\right)\right] \cdot \delta t$$

where there is a double summation giving nine terms in the major brackets.]

Since H is the internal heat generated per unit volume per unit time, the heat generated for the elementary volume in time δt is

$$\delta Q_{\text{gen}} = H(\delta x \delta y \delta z) \cdot \delta t$$

The net increase of internal energy of the volume in time δt because of δQ_{flow} and δQ_{gen} is, from eqn. (2.1.11),

$$\delta Q_{\text{inc}} = \rho(\delta x \delta y \delta z)C_{\text{p}} \cdot \frac{\partial T}{\partial t} \cdot \delta t$$

The conservation of energy principle applied to the element is

$$\delta Q_{\text{flow}} + \delta Q_{\text{gen}} = \delta Q_{\text{inc}}$$

which, upon substitution and division by $(\delta x \delta y \delta z) \cdot \delta t$, becomes

(i) $$k\left(\frac{\partial^2 T}{\partial x^2} + \frac{\partial^2 T}{\partial y^2} + \frac{\partial^2 T}{\partial z^2}\right) + H = \rho C_{\text{p}} \frac{\partial T}{\partial t} \qquad (2.2.4)$$

(ii) $$\frac{\partial}{\partial x}\left(k\frac{\partial T}{\partial x}\right) + \frac{\partial}{\partial y}\left(k\frac{\partial T}{\partial y}\right) + \frac{\partial}{\partial z}\left(k\frac{\partial T}{\partial z}\right) + H = \rho C_{\text{p}} \frac{\partial T}{\partial t}$$

where

$$k = k(x, y, z, T, t)$$

(iii) $$\frac{\partial}{\partial x}\left(k_x\frac{\partial T}{\partial x}\right) + \frac{\partial}{\partial y}\left(k_y\frac{\partial T}{\partial y}\right) + \frac{\partial}{\partial z}\left(k_z\frac{\partial T}{\partial z}\right) + H = \rho C_{\text{p}} \frac{\partial T}{\partial t}$$

(iv) $$\frac{\partial}{\partial x}\left(k_{1j}\frac{\partial T}{\partial x_j}\right) + \frac{\partial}{\partial y}\left(k_{2j}\frac{\partial T}{\partial x_j}\right) + \frac{\partial}{\partial z}\left(k_{3j}\frac{\partial T}{\partial x_j}\right) + H = \rho C_{\text{p}} \frac{\partial T}{\partial t}$$

[Again, the last one of these may also be written

$$\frac{\partial}{\partial x_i}\left(k_{ij}\frac{\partial T}{\partial x_j}\right) + H = \rho C_{\text{p}} \frac{\partial T}{\partial t}\ ]$$

This is the general equation of heat conduction which may also be written, in the case of (i) and (ii), as

(i) $$k\,\nabla^2 T + H = \rho C_{\text{p}} \frac{\partial T}{\partial t} \qquad (2.2.5)$$

(ii) $$\nabla \cdot (k\,\nabla T) + H = \rho C_{\text{p}} \frac{\partial T}{\partial t}$$

2.3. GENERAL EQUATION IN OTHER COORDINATE SYSTEMS

The equations developed in Section 2.2 are in terms of a cartesian coordinate system (x, y, z). Often, practical problems involve cylindrical or spherical symmetry and boundary conditions are often given on a surface which possess the property of one coordinate being held constant. In these cases the cylindrical polar coordinate system (r, θ, z) and the spherical polar coordinate system (r, θ, ϕ) are appropriate (see Figs. 2.4 and 2.5).

Sometimes other coordinate systems are appropriate and these will be mentioned briefly. In order to prepare the way for the solution of problems in cylindrically or spherically symmetric regions, a short account of these two coordinate systems is given, together with the appropriate forms of the general equation. In this connection it should be noted that it is only the Laplacian term $\nabla^2 T$ which changes, and for simplicity the constant

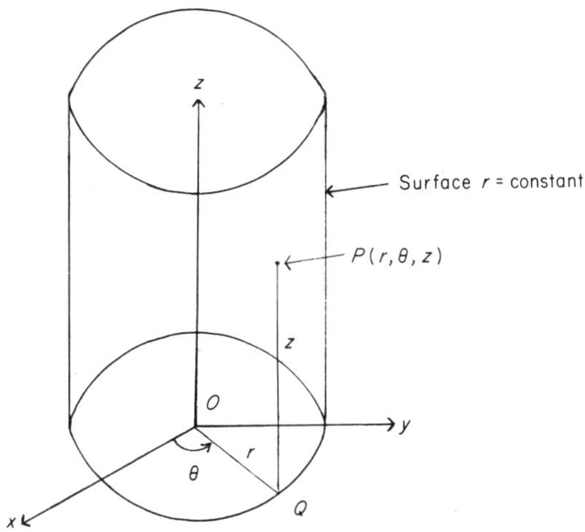

FIG. 2.4. Cylindrical polar coordinate system.

conductivity version (equations labelled with the (i) in Section 2.2) will be treated here, it being a simple task to write down the other versions from the final forms.

First, consider the cylindrical polar system shown in Fig. 2.4. Let P be a typical point in space whose projection on the xy-plane is the point Q. Then if $r = OQ$, $z = QP$, and θ is the angle made by OQ with the positive x-axis, the point P is uniquely specified by the three coordinates (r, θ, z). These are the cylindrical polar coordinates of P, and are related to the rectangular cartesian coordinates (x, y, z) of P by the equations

$$x = r\cos\theta; \quad y = r\sin\theta; \quad z = z \tag{2.3.1}$$

From these equations it follows directly that

$$r = (x^2 + y^2)^{1/2}; \quad \theta = \tan^{-1}(y/x) \tag{2.3.2}$$

Hence, if T is a function of x, y and z (and therefore a function of r, θ, z also), it follows from the elementary rules of partial differentiation that

$$\frac{\partial T}{\partial x} = \frac{\partial T}{\partial r}\frac{\partial r}{\partial x} + \frac{\partial T}{\partial \theta}\frac{\partial \theta}{\partial x} + \frac{\partial T}{\partial z}\frac{\partial z}{\partial x}$$

$$= \frac{x}{(x^2+y^2)^{1/2}}\frac{\partial T}{\partial r} - \frac{y}{x^2+y^2}\frac{\partial T}{\partial \theta}$$

$$= \cos\theta\frac{\partial T}{\partial r} - \frac{\sin\theta}{r}\frac{\partial T}{\partial \theta}$$

Consequently the operator

$$\frac{\partial}{\partial x} = \cos\theta\frac{\partial}{\partial r} - \frac{\sin\theta}{r}\frac{\partial}{\partial \theta}$$

and hence

$$\frac{\partial^2 T}{\partial x^2} = \frac{\partial}{\partial x}\left(\frac{\partial T}{\partial x}\right) = \left(\cos\theta\frac{\partial}{\partial r} - \frac{\sin\theta}{r}\frac{\partial}{\partial \theta}\right)\left(\cos\theta\frac{\partial T}{\partial r} - \frac{\sin\theta}{r}\frac{\partial T}{\partial \theta}\right)$$

$$= \cos^2\theta\frac{\partial^2 T}{\partial r^2} - \frac{2\sin\theta\cos\theta}{r}\frac{\partial^2 T}{\partial r\,\partial\theta} + \frac{\sin^2\theta}{r^2}\frac{\partial^2 T}{\partial \theta^2}$$

$$+ \frac{\sin^2\theta}{r}\frac{\partial T}{\partial r} + \frac{2\sin\theta\cos\theta}{r^2}\frac{\partial T}{\partial \theta} \qquad (2.3.3)$$

Similarly, using eqns. (2.3.1) and (2.3.2):

$$\frac{\partial T}{\partial y} = \frac{\partial T}{\partial r}\frac{\partial r}{\partial y} + \frac{\partial T}{\partial \theta}\frac{\partial \theta}{\partial y} + \frac{\partial T}{\partial z}\frac{\partial z}{\partial y} = \sin\theta\frac{\partial T}{\partial r} + \frac{\cos\theta}{r}\frac{\partial T}{\partial \theta}$$

and hence

$$\frac{\partial^2 T}{\partial y^2} = \frac{\partial}{\partial y}\left(\frac{\partial T}{\partial y}\right) = \left(\sin\theta\frac{\partial}{\partial r} + \frac{\cos\theta}{r}\frac{\partial}{\partial \theta}\right)\left(\sin\theta\frac{\partial T}{\partial r} + \frac{\cos\theta}{r}\frac{\partial T}{\partial \theta}\right)$$

$$= \sin^2\theta\frac{\partial^2 T}{\partial r^2} + \frac{2\sin\theta\cos\theta}{r}\frac{\partial^2 T}{\partial r\,\partial\theta} + \frac{\cos^2\theta}{r^2}\frac{\partial^2 T}{\partial \theta^2}$$

$$+ \frac{\cos^2\theta}{r}\frac{\partial T}{\partial r} - \frac{2\sin\theta\cos\theta}{r^2}\frac{\partial T}{\partial \theta} \qquad (2.3.4)$$

Hence, using eqns. (2.3.3) and (2.3.4):

$$\nabla^2 T \equiv \frac{\partial^2 T}{\partial x^2} + \frac{\partial^2 T}{\partial y^2} + \frac{\partial^2 T}{\partial z^2}$$

$$= \frac{\partial^2 T}{\partial r^2} + \frac{1}{r}\frac{\partial T}{\partial r} + \frac{1}{r^2}\frac{\partial^2 T}{\partial \theta^2} + \frac{\partial^2 T}{\partial z^2}$$

which is the Laplacian of T in cylindrical polar coordinates. An equivalent form is

$$\nabla^2 T = \frac{1}{r}\frac{\partial}{\partial r}\left(r\frac{\partial T}{\partial r}\right) + \frac{1}{r^2}\frac{\partial^2 T}{\partial \theta^2} + \frac{\partial^2 T}{\partial z^2}$$

Hence the general equation with constant conductivity can be written

$$k\left[\frac{1}{r}\frac{\partial}{\partial r}\left(r\frac{\partial T}{\partial r}\right) + \frac{1}{r^2}\frac{\partial^2 T}{\partial \theta^2} + \frac{\partial^2 T}{\partial z^2}\right] + H = \rho C_{\mathrm{p}}\frac{\partial T}{\partial t} \qquad (2.3.5)$$

and the form for variable conductivity is readily deducible as

$$\frac{1}{r}\frac{\partial}{\partial r}\left(rk\frac{\partial T}{\partial r}\right) + \frac{1}{r^2}\frac{\partial}{\partial \theta}\left(k\frac{\partial T}{\partial \theta}\right) + \frac{\partial}{\partial z}\left(k\frac{\partial T}{\partial z}\right) + H = \rho C_{\mathrm{p}}\frac{\partial T}{\partial t} \qquad (2.3.6)$$

Similar forms are appropriate for other assumptions about k.

Now consider the spherical polar coordinate system as illustrated in Fig. 2.5. Again let P be a typical point space whose projection on the xy-plane is

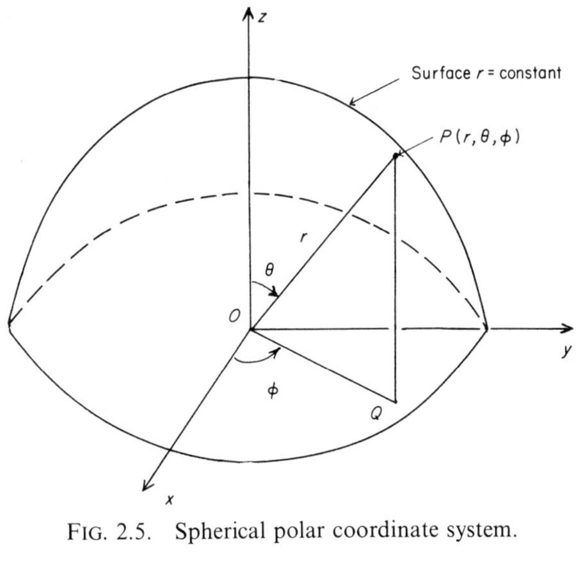

FIG. 2.5. Spherical polar coordinate system.

the point Q. Then if $r = OP$, ϕ the angle made by OQ with the positive x-axis, and θ the angle made by OP with the positive z-axis, the point P is uniquely specified by the three coordinates (r, ϕ, θ). These are the spherical polar coordinates of P, and are related to its rectangular cartesian coordinates by the relations

$$x = r\sin\theta\cos\phi; \quad y = r\sin\theta\sin\phi; \quad z = r\cos\theta$$

Following a similar analysis to that given above for cylindrical polar coordinates, it can be shown that in spherical polar coordinates

$$\nabla^2 T = \frac{\partial^2 T}{\partial r^2} + \frac{2}{r}\frac{\partial T}{\partial r} + \frac{1}{r^2 \sin^2\theta}\frac{\partial^2 T}{\partial \phi^2} + \frac{1}{r^2}\frac{\partial^2 T}{\partial \theta^2} + \frac{\cot\theta}{r^2}\frac{\partial T}{\partial \theta}$$

$$= \frac{1}{r^2}\left[\frac{\partial}{\partial r}\left(r^2\frac{\partial T}{\partial r}\right) + \frac{1}{\sin^2\theta}\frac{\partial^2 T}{\partial \phi^2} + \frac{1}{\sin\theta}\frac{\partial}{\partial \theta}\left(\sin\theta\frac{\partial T}{\partial \theta}\right)\right]$$

Thus the general equation with constant conductivity may be written as

$$\frac{k}{r^2}\left[\frac{\partial}{\partial r}\left(r^2\frac{\partial T}{\partial r}\right) + \frac{1}{\sin^2\theta}\frac{\partial^2 T}{\partial \phi^2} + \frac{1}{\sin\theta}\frac{\partial}{\partial \theta}\left(\sin\theta\frac{\partial T}{\partial \theta}\right)\right] + H = \rho C_{\mathrm{p}}\frac{\partial T}{\partial t} \quad (2.3.7)$$

and the form for variable k as

$$\frac{1}{r^2}\left[\frac{\partial}{\partial r}\left(r^2 k\frac{\partial T}{\partial r}\right) + \frac{1}{\sin^2\theta}\frac{\partial}{\partial \phi}\left(k\frac{\partial T}{\partial \phi}\right) + \frac{1}{\sin\theta}\frac{\partial}{\partial \theta}\left(k\sin\theta\frac{\partial T}{\partial \theta}\right)\right] + H = \rho C_{\mathrm{p}}\frac{\partial T}{\partial t} \quad (2.3.8)$$

Other orthogonal coordinate systems can be handled by knowledge of the form taken by $\nabla^2 T$ is given in each case. This term is the scalar Laplacian and may be evaluated in an orthogonal coordinate system from

$$\nabla^2 T = \frac{1}{h_1 h_2 h_3}\left[\frac{\partial}{\partial x_1}\left(\frac{h_2 h_3}{h_1}\frac{\partial T}{\partial x_1}\right) + \frac{\partial}{\partial x_2}\left(\frac{h_1 h_3}{h_2}\frac{\partial T}{\partial x_2}\right) + \frac{\partial}{\partial x_3}\left(\frac{h_1 h_2}{h_3}\frac{\partial T}{\partial x_3}\right)\right] \quad (2.3.9)$$

where the h_1, h_2 and h_3 are the metric coefficients of the coordinate system. Thus knowledge of these and the relationships between the coordinates and the cartesian coordinates x, y and z enables this expression to be written down as required. Table 2.3 lists the required details for the cartesian,

TABLE 2.3

COORDINATE RELATIONSHIPS AND METRIC COEFFICIENTS FOR CARTESIAN, CYLINDRICAL AND SPHERICAL COORDINATE SYSTEMS

	Cartesian	*Cylindrical* $x = r\cos\theta$ $y = r\sin\theta$ $z = z$	*Spherical* $x = r\sin\theta\cos\phi$ $y = r\sin\theta\sin\phi$ $z = r\cos\theta$
x_1	x	r	r
h_1^2	1	1	1
x_2	y	θ	ϕ
h_2^2	1	r^2	$r^2\sin^2\theta$
x_3	z	z	θ
h_3^2	1	1	r^2

cylindrical and spherical coordinate systems; details regarding many other systems are available in Happel & Brenner (1965).

2.4. SPECIAL FORMS OF THE GENERAL EQUATION

The general equations of Sections 2.1–2.3 represent the first step in obtaining the solution of a heat transfer problem. The general equation, in the appropriate coordinate system and with appropriate conductivity assumptions, is reduced to its simplest form consistent with the conditions established by the problem. There is no need here to elaborate further on the diffusion ($H = 0$), Poisson ($\partial T/\partial t = 0$) or Laplace ($H = 0$ and $\partial T/\partial t = 0$) versions (see Section 1.3 of Chapter 1) of the general equation. However, one can often remove a non-linearity or directional conductivity and sometimes the problem is not fully 3-D and time-dependent, in which case several of the terms may be omitted.

2.4.1. Non-linearity Removal

The equation (eqn. (2.2.5))

$$\nabla \cdot (k \nabla T) + H = \rho C_{\mathrm{p}} \frac{\partial T}{\partial t} \tag{2.4.1}$$

which governs the heat transfer in an isotropic material with conductivity dependent on temperature $k = k(T)$ possesses a non-linearity in its first term. In some problems this may be removed by the Kirchhoff transformation, where the dependent variable T is transformed to a new variable V defined by

$$\mathrm{d}V = \frac{k(T)}{k_0} \cdot \mathrm{d}T = \frac{k}{k_0} \mathrm{d}T$$

where $k_0 = k(T_0)$. Typical derivatives of eqn. (2.4.1) are then written in terms of V by way of equations like

$$\frac{\partial V}{\partial t} = \frac{\partial V}{\partial T} \cdot \frac{\partial T}{\partial t} = \frac{k}{k_0} \frac{\partial T}{\partial t}; \quad \frac{\partial^2 V}{\partial x^2} = \frac{1}{k_0} \frac{\partial}{\partial x}\left(k \frac{\partial T}{\partial x}\right)$$

and eqn. (2.4.1) now reduces to

$$\nabla^2 V + \frac{H}{k_0} = \frac{\rho C_{\mathrm{p}}}{k} \frac{\partial V}{\partial t} = \frac{1}{\alpha} \frac{\partial V}{\partial t}$$

This equation is now linear in V for steady-state problems where the term on the right is omitted. It is linear in V for unsteady-state conditions, however, only if C_p and k are always proportional to each other. Even in cases when this is not so, it is often simpler and more accurate to treat the equation in this form than in its original non-linear form.

2.4.2. Directional Conductivity Removal

Another interesting transformation removes the problem of constant directional conductivities k_x, k_y and k_z for problems of orthotropic materials (see equations labelled (iii) in Section 2.2). The x, y and z coordinates are scaled in terms of new space coordinates X, Y and Z via

$$X = \left(\frac{k_0}{k_x}\right)^{1/2} x; \quad Y = \left(\frac{k_0}{k_y}\right)^{1/2} y; \quad Z = \left(\frac{k_0}{k_z}\right)^{1/2} z$$

where k_0 is any fixed value and the equation

$$\frac{\partial}{\partial x}\left(k_x \frac{\partial T}{\partial x}\right) + \frac{\partial}{\partial y}\left(k_y \frac{\partial T}{\partial y}\right) + \frac{\partial}{\partial z}\left(k_z \frac{\partial T}{\partial z}\right) + H = \rho C_p \frac{\partial T}{\partial t}$$

reduces to the simplified form

$$\frac{\partial^2 T}{\partial x^2} + \frac{\partial^2 T}{\partial y^2} + \frac{\partial^2 T}{\partial z^2} + \frac{H}{k_0} = \frac{\rho C_p}{k_0} \frac{\partial T}{\partial t}$$

Later this process will be seen to be equivalent to using different mesh spacings in the three directions when the same FDE results for a typical point.

2.4.3. Reduced Forms of the General Equation

In the case of problems which do not have three space coordinates and time as independent variables, several terms of the governing equations may be omitted. For simplicity, the constant conductivity forms in the cartesian, cylindrical and spherical coordinate systems are considered. Tables 2.4 to 2.6 present some of these simplified versions of the equations; later the form of FDE in each case is given (see Tables 4.1 to 4.3 of Chapter 4).

2.5. PROBLEM CLASSIFICATION

The general form of the partial differential equation for $T = T(x_1, x_2, \ldots, x_n)$ is

$$\sum_{i=1}^{n} A_i \frac{\partial^2 T}{\partial x_i^2} + \sum_{i=1}^{n} B_i \frac{\partial T}{\partial x_i} + CT + D = 0 \qquad (2.5.1)$$

TABLE 2.4

REDUCED FORM OF GENERAL EQUATION WITH CONSTANT CONDUCTIVITY IN CARTESIAN COORDINATES

Problem	*Heat flow equation*	*Comments*
2-D steady-state	$\frac{\partial^2 T}{\partial x^2} + \frac{\partial^2 T}{\partial y^2} = 0$	$H = 0,\ \frac{\partial T}{\partial t} = 0,\ \frac{\partial T}{\partial z} = 0$
3-D steady-state with internal heat generation	$\frac{\partial^2 T}{\partial x^2} + \frac{\partial^2 T}{\partial y^2} + \frac{\partial^2 T}{\partial z^2} + \frac{H}{k} = 0$	$\frac{\partial T}{\partial t} = 0$
1-D transient	$\alpha\left(\frac{\partial^2 T}{\partial x^2}\right) = \frac{\partial T}{\partial t}$	$H = 0,\ \frac{\partial T}{\partial y} = \frac{\partial T}{\partial z} = 0$
2-D transient with internal heat generation	$\frac{H}{\rho C_p} + \alpha\left(\frac{\partial^2 T}{\partial x^2} + \frac{\partial^2 T}{\partial y^2}\right) = \frac{\partial T}{\partial t}$	$\frac{\partial T}{\partial z} = 0$

TABLE 2.5

REDUCED FORM OF GENERAL EQUATION WITH CONSTANT CONDUCTIVITY IN CYLINDRICAL POLAR COORDINATES

Problem	*Heat flow equation*	*Comments*
2-D steady-state Disk	$\frac{1}{r}\frac{\partial}{\partial r}\left(r\frac{\partial T}{\partial r}\right)+\frac{1}{r^2}\frac{\partial^2 T}{\partial \theta^2}=0$	$\frac{\partial T}{\partial z}=0$; $H=0$, $\frac{\partial T}{\partial t}=0$
Cylinder	$\frac{1}{r}\frac{\partial}{\partial r}\left(r\frac{\partial T}{\partial r}\right)+\frac{\partial^2 T}{\partial z^2}=0$	$\frac{\partial T}{\partial \theta}=0$; $H=0$, $\frac{\partial T}{\partial t}=0$
3-D steady-state with internal heat generation	$\frac{1}{r}\frac{\partial}{\partial r}\left(r\frac{\partial T}{\partial r}\right)+\frac{1}{r^2}\frac{\partial^2 T}{\partial \theta^2}+\frac{\partial^2 T}{\partial z^2}+\frac{H}{k}=0$	$\frac{\partial T}{\partial t}=0$
1-D transient	$\alpha\cdot\frac{1}{r}\frac{\partial}{\partial r}\left(r\frac{\partial T}{\partial r}\right)=\frac{\partial T}{\partial t}$	$H=0,\quad \frac{\partial T}{\partial \theta}=\frac{\partial T}{\partial z}=0$
2-D transient with internal heat generation Disk	$\frac{H}{\rho C_p}+\alpha\left[\frac{1}{r}\frac{\partial}{\partial r}\left(r\frac{\partial T}{\partial r}\right)+\frac{1}{r^2}\frac{\partial^2 T}{\partial \theta^2}\right]=\frac{\partial T}{\partial t}$	$\frac{\partial T}{\partial z}=0$
Cylinder	$\frac{H}{\rho C_p}+\alpha\left[\frac{1}{r}\frac{\partial}{\partial r}\left(r\frac{\partial T}{\partial r}\right)+\frac{\partial^2 T}{\partial z^2}\right]=\frac{\partial T}{\partial t}$	$\frac{\partial T}{\partial \theta}=0$

TABLE 2.6
REDUCED FORM OF GENERAL EQUATION WITH CONSTANT CONDUCTIVITY IN SPHERICAL POLAR COORDINATES

Problem	*Heat flow equation*	*Comments*
2-D steady-state	$\frac{\partial}{\partial r}\left(r^2\frac{\partial T}{\partial r}\right)+\frac{1}{\sin\theta}\frac{\partial}{\partial\theta}\left(\sin\theta\frac{\partial T}{\partial\theta}\right)=0$	$\frac{\partial T}{\partial\phi}=0$ longitudinal symmetry; $H=0$, $\frac{\partial T}{\partial t}=0$
	$\frac{\partial}{\partial r}\left(r^2\frac{\partial T}{\partial r}\right)+\frac{1}{\sin^2\theta}\frac{\partial^2 T}{\partial\phi^2}=0$	$\frac{\partial T}{\partial\theta}=0$ latitudinal symmetry; $H=0$, $\frac{\partial T}{\partial t}=0$
3-D steady-state with internal heat generation	$\frac{\partial}{\partial r}\left(r^2\frac{\partial T}{\partial r}\right)+\frac{1}{\sin^2\theta}\frac{\partial^2 T}{\partial\phi^2}+\frac{1}{\sin\theta}\frac{\partial}{\partial\theta}\left(\sin\theta\frac{\partial T}{\partial\theta}\right)+\frac{Hr^2}{k}=0$	$\frac{\partial T}{\partial t}=0$
1-D transient	$\frac{\alpha}{r^2}\frac{\partial}{\partial r}\left(r^2\frac{\partial T}{\partial r}\right)=\frac{\partial T}{\partial t}$	$H=0,\ \frac{\partial T}{\partial\theta}=\frac{\partial T}{\partial\phi}=0$
2-D transient with internal heat generation	$\frac{H}{\rho C_p}+\frac{\alpha}{r^2}\left[\frac{\partial}{\partial r}\left(r^2\frac{\partial T}{\partial r}\right)+\frac{1}{\sin\theta}\frac{\partial}{\partial\theta}\left(\sin\theta\frac{\partial T}{\partial\theta}\right)\right]=\frac{\partial T}{\partial t}$	$\frac{\partial T}{\partial\phi}=0$ longitudinal symmetry
	$\frac{H}{\rho C_p}+\frac{\alpha}{r^2}\left[\frac{\partial}{\partial r}\left(r^2\frac{\partial T}{\partial r}\right)+\frac{1}{\sin^2\theta}\frac{\partial^2 T}{\partial\phi^2}\right]=\frac{\partial T}{\partial t}$	$\frac{\partial T}{\partial\theta}=0$ latitudinal symmetry

where the x_i's may be any of three space directions (x, y, z in the cartesian system, for example) or time t. Summation is over the number of independent variables n; for a one space dimension and time problem with the governing equation

$$\frac{\partial T}{\partial t} = \alpha \frac{\partial^2 T}{\partial x^2}$$

we have $n = 2$, $x_1 = x$ and $x_2 = t$. This general equation, eqn. (2.5.1), is classified as:

(i) *Hyperbolic* if all the A_i's are non-zero and have with one exception the same sign, for example

$$\frac{\partial^2 T}{\partial x^2} - \alpha \frac{\partial^2 T}{\partial t^2} = 0 \quad \text{(wave equation)}$$

(ii) *Elliptic* if all the A_i's are non-zero and have the same sign, for example

$$\frac{\partial^2 T}{\partial x^2} + \frac{\partial^2 T}{\partial y^2} = 0$$

(steady-state Laplace equation in two space dimensions x and y). Boundary conditions are required on the whole of the outer boundary of the region in which the solution is required—either conditions on T itself (Dirichlet conditions) or on its normal outward derivative $\partial T/\partial n$ (Neumann conditions), but T itself must be specified on at least part of the boundary. These are illustrated in Fig. 2.6. Other elliptic equations are

$$\frac{\partial^2 T}{\partial x^2} = 0 \quad \text{(1-D steady-state)}$$

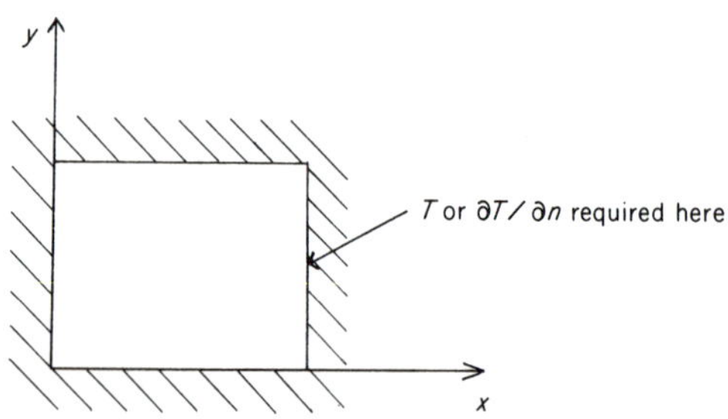

FIG. 2.6. Boundary conditions required for 2-D elliptic PDE.

and

$$\frac{\partial^2 T}{\partial x^2} + \frac{\partial^2 T}{\partial y^2} + \frac{\partial^2 T}{\partial z^2} = 0 \quad \text{(3-D steady-state)}$$

and similar equations with an internal heat generation term, like

$$k\frac{\partial^2 T}{\partial x^2} + H = 0 \quad \text{(1-D steady-state with heat generation)}$$

(iii) *Parabolic* if one A_i is zero (A_k for instance) and the remaining A_i's are non-zero and have the same sign and the coefficient B_k is non-zero, for example

$$\frac{\partial T}{\partial t} - \alpha\left(\frac{\partial^2 T}{\partial x^2} + \frac{\partial^2 T}{\partial y^2}\right) = 0$$

(transient equation in two space dimensions and time). For heat transfer problems usually $x_k = t$. Boundary conditions are required for T itself on the initial line or region at $t = 0$ together with T (Dirichlet) or $\partial T/\partial n$ (Neumann) on the edges of the region for $t > 0$. These are illustrated in Fig. 2.7. Because of these boundary condition requirements, these problems are sometimes known as initial value problems.

Other parabolic equations are the obvious ones representing transient heat flow in one or three space dimensions and time; again these may include an internal heat generation term.

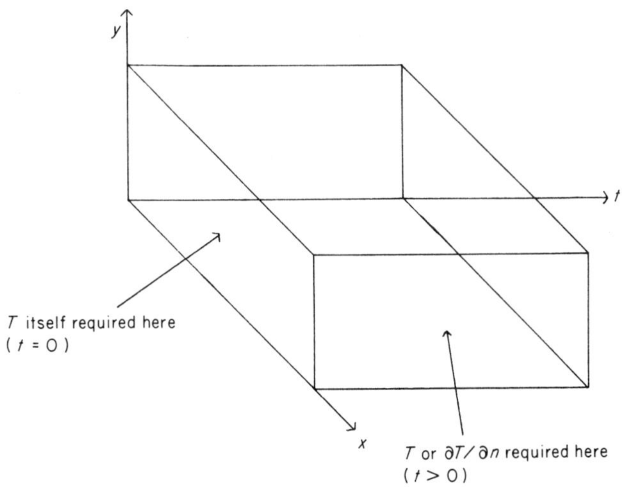

FIG. 2.7. Boundary conditions required for 2-D space plus time parabolic PDE.

BIBLIOGRAPHY

ABRAMOWITZ, M. & STEGUN, I. A. (eds) (1964). *Handbook of Mathematical Functions*. Applied Mathematics Series No. 55, National Bureau of Standards, US Dept of Commerce, Washington, D.C.

BAYLEY, F. J., OWEN, J. M. & TURNER, A. B. (1972). *Heat Transfer*. Nelson, London.

GEBHART, B. (1971). *Heat Transfer*, 2nd ed. McGraw-Hill, New York.

GRAY, W. A. & MÜLLER, R. (1974). *Engineering Calculations in Radiative Heat Transfer*. Pergamon Press, Oxford.

HAPPEL, J. & BRENNER, H. (1965). *Low Reynolds Number Hydrodynamics with Special Applications to Particulate Media*. Prentice-Hall, Englewood Cliffs, N.J.

MOON, P. & SPENCER, D. E. (1965). *Vectors*. Van Nostrand, Princeton, N.J.

OPEN UNIVERSITY (1973). *Partial Differential Equations of Applied Mathematics: A Third Level Mathematics Course*. The Open University Press, Walton Hall, Milton Keynes, Bucks., UK.

RAZELOS, P. (1973). Methods of obtaining approximate solutions. Section 4, pp. 4–1 to 4–78, in *Handbook of Heat Transfer*, ed. W. M. Rohsenow and J. P. Hartnett. McGraw-Hill, New York.

ROGERS, G. & MAYHEW, Y. (1967). *Engineering Thermodynamics, Work and Heat Transfer*. Longmans, London.

SCHNEIDER, P. J. (1973). Conduction. Section 3, pp. 3–1 to 3–134, in *Handbook of Heat Transfer*, ed. W. M. Rohsenow and J. P. Hartnett. McGraw-Hill, New York.

SIMONSON, R. J. (1962). *An introduction to Engineering Heat Transfer*. McGraw-Hill, New York.

STEPHENSON, G. (1970). *An Introduction to Partial Differential Equations for Science Students*, 2nd ed. Longman, London.

CHAPTER 3

Basics of Finite Difference Equations in Cartesian Coordinates

3.1. FUNCTION DERIVATIVES IN UNIFORM GRID

Fundamental to the numerical solution of PDEs are the expressions for derivatives of a function $f(x)$ at some point in terms of function values at several neighbouring points. The point may be to one side of the neighbours, or it may lie within them. The points may be positioned with equal spacing Δx or the spacing may be irregular. One may be particularly interested in the truncation error (the likely accuracy of the formula used) and wish to know precisely the order of accuracy of the formula being used, or alternatively one may be guided by experience and intuition, and prefer to enjoy the simplicity offered by some of the methods. For completeness of this section, several of the common techniques used widely are described, and further information is available in Carnahan *et al.* (1969) and Gerald (1970). Here one is constantly interested in what is the 'best' formula for approximating $f'(x_0)$ (say) in terms of local points, from the point of view of accuracy of the formula and convenience in subsequent use. The typical example might be as shown in Fig. 3.1. In this section base-point spacing is restricted to be uniform and equal to a fixed Δx; the next section allows non-uniformities.

3.1.1. Heat Balance

It is worth noting immediately that when the heat balance approach is used to set up the simulation of the problem, there is no need to deal with the governing PDE. Direct approximation to the laws governing the physical system is applied locally to each cell control volume surrounding each grid point. The engineer wishing to use this technique exclusively will have little need for the other techniques whose bases are now to be described, and he

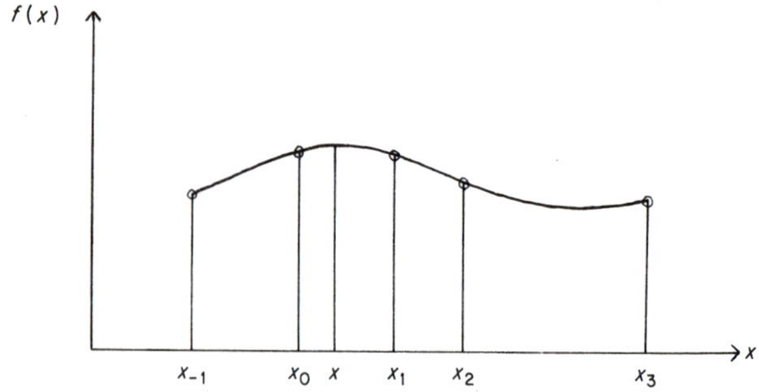

FIG. 3.1. Typical numerical differentiation problem: what is the best approximation to $f'(x)$ and $f''(x)$ in terms of function values at x_{-1}, x_1, x_2, x_3?

need read little more of this section. He need approximate only first derivatives, and intuition will guide him as described below. The details are illustrated later in Sections 3.4 and 3.5.2.

3.1.2. Intuition

Consider the function $f(x)$ shown in Fig. 3.2. The notation is as follows: x_{-1}, x_0 and x_1 are equally spaced base-points with $x_1 - x_0 = x_0 - x_{-1} = \Delta x$, that is, $x_i = x_0 + i\Delta x$, and function values are denoted by $f(x)$ and $f_i = f(x)|_{x=x_i}$. Intermediate values are used denoted by an appropriate non-integer value being given to x, for example

$$f_{-1/2} = f(x)|_{x=x_{-1/2}} = f(x_{-1/2}) = f(x_0 + (-\tfrac{1}{2}) \cdot \Delta x)$$

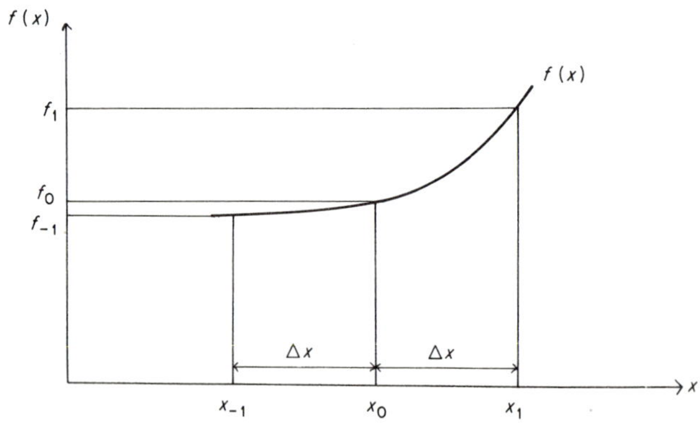

FIG. 3.2. Typical $f(x)$ showing function values f_{-1}, f_0 and f_1 taken at equally spaced base-points x_{-1}, x_0, x_1 distant Δx between successive points.

The intuitive estimate of f' at $x = x_{-1/2}$ is

$$\left.\frac{\mathrm{d}f}{\mathrm{d}x}\right|_{x=x_{-1/2}} = \frac{f_0 - f_{-1}}{\Delta x} \tag{3.1.1}$$

based on f_{-1} and f_0 only, and at $x = x_{1/2}$ it is

$$\left.\frac{\mathrm{d}f}{\mathrm{d}x}\right|_{x=x_{1/2}} = \frac{f_1 - f_0}{\Delta x} \tag{3.1.2}$$

based on f_0 and f_1 only. If we infer that either of these is an approximation to $\mathrm{d}f/\mathrm{d}x$ at $x = x_0$, they are called one-sided differences, the first one being a backward difference and the second one a forward difference. Clearly, a better estimate at $x = x_0$ would be made using the two points on either side of x_0 only:

$$\left.\frac{\mathrm{d}f}{\mathrm{d}x}\right|_{x=x_0} = \frac{f_1 - f_{-1}}{2\Delta x} \tag{3.1.3}$$

which is called the central difference estimate based on the two values f_1 and f_{-1}. It is in fact equal to the average of the two one-sided estimates. With the operator notation

$$\begin{aligned} \Delta f_0 &= f_1 - f_0 && \text{(forward)} \\ \nabla f_0 &= f_0 - f_{-1} && \text{(backward)} \\ \delta f_0 &= f_{1/2} - f_{-1/2} && \text{(central)} \\ \mu\phi &= \tfrac{1}{2}(\phi_{1/2} + \phi_{-1/2}) && \text{(averaging)} \end{aligned}$$

the numerators in the eqns. (3.1.1)–(3.1.3) may be represented by ∇f_0, Δf_0 and $\mu\delta f_0$ respectively. Clearly, the central first-order derivative formula

$$\left.\frac{\mathrm{d}f}{\mathrm{d}x}\right|_{x=x_0} = \frac{f_{1/2} - f_{-1/2}}{\Delta x}$$

whose numerator is δf_0, is probably a better estimate than any of the three but it involves non-tabular points. We find later, as expected, that the central difference formula $\mu\delta f_0/(2\Delta x)$ is more accurate than either of the two one-sided formulae $\Delta f_0/\Delta x$ and $\nabla f_0/\Delta x$. All three of the above estimates for $f'(x_0)$ are clearly approximations, and as Δx becomes smaller the estimates improve.

An estimate of the second derivative $f''(x)$ at $x = x_0$ is obtained by estimating the first derivative in each of the two intervals (x_{-1}, x_0) and

(x_0, x_1) and dividing their difference by Δx, the distance between the two points where these estimates apply. Hence

$$\left.\frac{d^2 f}{dx^2}\right|_{x=x_0} = \frac{d}{dx}\left(\frac{df}{dx}\right)\bigg|_{x=x_0}$$

$$= \frac{\left.\dfrac{df}{dx}\right|_{x=x_{1/2}} - \left.\dfrac{df}{dx}\right|_{x=x_{-1/2}}}{\Delta x}$$

$$= \frac{\dfrac{(f_1 - f_0)}{\Delta x} - \dfrac{(f_0 - f_{-1})}{\Delta x}}{\Delta x}$$

$$= \frac{f_1 - 2f_0 + f_{-1}}{(\Delta x)^2}$$

Again accuracy improves as Δx becomes smaller.

This and other formulae for the second-order derivative could have been obtained readily as follows:

$$\Delta^2 f_0 = \Delta(\Delta f_0) = \Delta f_1 - \Delta f_0 = f_2 - 2f_1 + f_0$$

$$\nabla^2 f_0 = \nabla(\nabla f_0) = \nabla f_0 - \nabla f_{-1} = f_0 - 2f_{-1} + f_{-2}$$

$$\delta^2 f_0 = \delta(\delta f_0) = \delta f_{1/2} - \delta f_{-1/2} = f_1 - 2f_0 + f_1$$

from which division by $(\Delta x)^2$ yields forward, backward and central estimates of the second-order derivative at x_0. As expected, the last of these centred on x_0 turns out to be more accurate than the other two one-sided ones, which are actually centred at x_1 and x_{-1}.

3.1.3. Taylor Series

Whereas the above intuitive technique revealed plausible formulae for the first and second derivatives at $x = x_0$ based on values at two neighbours, x_{-1} on the left and x_1 on the right, no estimate of the error is possible. Neither is it possible to develop more accurate formulae by that technique. The use of Taylor series, however, has direct application to this.

First, consider the case of equally spaced base-points, as shown in Fig. 3.2, and recall the Taylor series expansion of $f(x_1) = f(x_0 + \Delta x)$ in terms of $\Delta x = x_1 - x_0$:

$$f_1 = f_0 + \Delta x f_0' + \frac{(\Delta x)^2}{2!} f_0'' + \frac{(\Delta x)^3}{3!} f_0''' + \frac{(\Delta x)^4}{4!} f_{0,1}^{iv}$$

when truncated after the term in $(\Delta x)^3$, the truncation error term being given where $f^{iv}_{0,1}$ stands for f^{iv} which is in the interval (x_0, x_1). Similarly the expansion of $f(x_{-1}) = f(x_0 - \Delta x)$ in terms of $\Delta x = x_0 - x_{-1}$ is

$$f_{-1} = f_0 - \Delta x f_0' + \frac{(\Delta x)^2}{2!} f_0'' - \frac{(\Delta x)^3}{3!} f_0''' + \frac{(\Delta x)^4}{4!} f^{iv}_{-1,0}$$

where $f^{iv}_{-1,0}$ is evaluated somewhere in (x_{-1}, x_0). Truncating after the term in Δx, these expansions give the forward- and backward-difference formulae for $f'(x_0)$ as

$$\left.\frac{df}{dx}\right|_{x=x_0} = \frac{f_1 - f_0}{\Delta x} - \frac{\Delta x}{2} f''_{0,1}$$

$$\left.\frac{df}{dx}\right|_{x=x_0} = \frac{f_0 - f_{-1}}{\Delta x} + \frac{\Delta x}{2} f''_{-1,0}$$

with error terms of order Δx and involving the second derivative of $f(x)$ evaluated somewhere in (x_0, x_1) and (x_{-1}, x_0) respectively. On the other hand, subtraction of these two expansions yields

$$\left.\frac{df}{dx}\right|_{x=x_0} = \frac{f_1 - f_{-1}}{2\Delta x} - \frac{(\Delta x)^2}{3!} (f'''_{-1,0} + f'''_{0,1})$$

with truncation error terms of order $(\Delta x)^2$ and involving the third derivative of $f(x)$ evaluated somewhere in (x_{-1}, x_1). This central difference formula is thus seen to be more accurate than the one-sided ones. Whilst it is true that this central difference formula has a second-order truncation error, this does not mean that its use will inevitably yield more accurate results than the one-sided formulae of first-order truncation error. In fact, it is well known that its application in a naïve way to the time derivative of an unsteady-state heat transfer equation always results in an unstable process, the results of which are useless, whereas the application of the forward-difference formula yields a satisfactory method of solution. To estimate the second derivative $f''(x)$ at $x = x_0$, addition of the two expansions reveals that

$$\left.\frac{d^2 f}{dx^2}\right|_{x=x_0} = \frac{f_1 - 2f_0 + f_{-1}}{(\Delta x)^2} - \frac{(\Delta x)^2}{4!} [f^{iv}_{0,1} + f^{iv}_{-1,0}]$$

where the error is of order $(\Delta x)^2$ and involves the fourth derivatives of $f(x)$ evaluated in (x_0, x_1) and (x_{-1}, x_0).

To obtain a more accurate representation of derivatives, one must use methods which can make use of more than three values of the function. The following two methods are to be recommended.

3.1.4. Interpolating Polynomials and Operators

In order to develop higher-order accuracy formulae, the use of higher-order interpolating polynomials passing through known function values is required. Since the formulae are based on polynomial approximation, they give exact results when operating on a polynomial of proper degree and it may be noted that there is a unique $(n - 1)$th degree polynomial in x passing through the n points $(x_i, f(x_i))$ $(1 \leq i \leq n)$, for example. When operating on an unknown function, there is an error which is usually represented by the largest of the omitted terms of the series form of the formula being used.

TABLE 3.1
FINITE DIFFERENCE OPERATORS

Forward difference:	$\Delta f_i = f_{i+1} - f_i$
Backward difference:	$\nabla f_i = f_i - f_{i-1}$
Central difference:	$\delta f_i = f_{i+1/2} - f_{i-1/2}$
Averaging	$\mu f_i = \frac{1}{2}(f_{i+1/2} + f_{i-1/2})$
Shift:	$Ef_i = f_{i+1}$
Differential:	$Df = \dfrac{\mathrm{d}f}{\mathrm{d}x}$

These higher-order formulae are most conveniently obtained by the use of symbolic finite difference operators defined, in the case of a $f(x)$ over a set of equally spaced base-points x_i with $h = \Delta x = x_{i+1} - x_i$, in Table 3.1.

These operators can usually be manipulated according to the usual laws of elementary algebra. The definitions lead immediately to the relations

$$\Delta = E - 1; \qquad \delta = E^{1/2} - E^{-1/2}$$
$$\nabla = 1 - E^{-1}; \quad \mu = \tfrac{1}{2}(E^{1/2} + E^{-1/2})$$

from which others may be obtained like

$$\Delta = \nabla E = \delta E^{1/2}; \qquad \mu^2 = 1 + \tfrac{1}{4}\delta^2$$

Taylor's theorem in the form

$$f(x + h) = f(x) + hf'(x) + \frac{h}{2!}f''(x) + \cdots$$

becomes

$$Ef(x) = \left(1 + hD + \frac{h^2D^2}{2!} + \cdots\right)f(x)$$

so that

$$E = \exp hD$$

All the interrelations between these operators are summarised in Table 3.2.

TABLE 3.2
RELATIONS BETWEEN SYMBOLIC OPERATORS

	E	Δ	δ	∇	hD
E	E	$1 + \Delta$	$1 + \frac{1}{2}\delta^2 + \delta(1 + \frac{1}{4}\delta^2)^{1/2}$	$(1 - \nabla)^{-1}$	$\exp hD$
Δ	$E - 1$	Δ	$\delta(1 + \frac{1}{4}\delta^2)^{1/2} + \frac{1}{2}\delta^2$	$\nabla(1 - \nabla)^{-1}$	$\exp(hD) - 1$
δ	$E^{1/2} - E^{-1/2}$	$\Delta(1 + \Delta)^{-1/2}$	δ	$\nabla(1 - \nabla)^{-1/2}$	$2 \sinh \frac{1}{2}hD$
∇	$1 - E^{-1}$	$\Delta(1 + \Delta)^{-1}$	$\delta(1 + \frac{1}{4}\delta^2)^{1/2} - \frac{1}{2}\delta^2$	∇	$1 - \exp(-hD)$
hD	$\log E$	$\log(1 + \Delta)$	$2 \sinh^{-1} \frac{1}{2}\delta$	$-\log(1 - \nabla)$	hD
μ	$\frac{1}{2}(E^{1/2} + E^{-1/2})$	$(1 + \frac{1}{2}\Delta)(1 + \Delta)^{-1/2}$	$(1 + \frac{1}{4}\delta^2)^{1/2}$	$(1 - \frac{1}{2}\nabla)(1 - \nabla)^{-1/2}$	$\cosh \frac{1}{2}hD$

Interpolating polynomials pass through some of the points $(x_i, f(x_i))$ (i integer) but they are also defined at intermediate values of x; their differentiation yields useful results. Formulae differ according to how many, and which, points are used. Newton–Gregory forward- and backward-difference interpolation formulae for $f_i = E^i f_0 = (1 + \Delta)^i f_0 = (1 - \nabla)^{-i} f_0$ may be written down by expanding $(1 + \Delta)^i$ and $(1 - \nabla)^{-i}$. These provide estimates of $f_i\ (i \geq 0)$ in terms of $f_0, f_1, f_2, \ldots$ and $f_0, f_{-1}, f_{-2}, \ldots$ respectively. Differentiating these formulae gives approximations for $f_i', f_i'', \ldots$ in terms of these ordinates. For the fine detail and further discussion of this material, the reader is recommended to use the publication by the National Physical Laboratory (1961).

3.1.5. Method of Undetermined Coefficients

A powerful alternative method of obtaining the required formulae for derivatives, albeit without much insight into the magnitude of the errors involved, is the method of undetermined coefficients (see Carnahan *et al.*, 1969; Gerald, 1970). Basically, certain conditions are imposed on a formula of desired form and these conditions suffice to determine values of the unknown coefficients in the formula. For example, a formula for the first derivative at x_0 may be suggested in terms of function values at the $n + 1$ equi-spaced base-points:

$$f'(x_0) = c_0 f_0 + c_1 f_1 + \cdots + c_n f_n$$

where $h = x_{i+1} - x_i$ for all i. Greater accuracy may be expected with the use of more points, since more information about the function is being used. The coefficients $c_0, c_1, \ldots, c_n$ are evaluated by requiring the formula to be exact whenever the function is a polynomial of degree n or less.

As an illustration, suppose one is looking for a first derivative formula

$$f_0' = c_0 f_0 + c_1 f_1$$

where f_0 and f_1 are known and c_0 and c_1 are unknown, and to be determined. The formula is to be exact whenever $f(x)$ is a polynomial of degree 1 or less, so it must hold if either $f(x) = x$ or $f(x) = 1$. Substituting for each of these cases yields the two equations:

$$\begin{array}{llll} f(x) = x & \text{and} \quad f'(x) = 1 & \text{so} & 1 = c_0(x_0) + c_1(x_0 + h) \\ f(x) = 1 & \text{and} \quad f'(x) = 0 & \text{so} & 0 = c_0(1) + c_1(1) \end{array}$$

Simultaneous solution yields $c_0 = -1/h$ and $c_1 = 1/h$ and so

$$f_0' \approx \frac{f_1 - f_0}{h}$$

Similarly, if looking for a three-term formula:

$$f_0' = c_0 f_0 + c_1 f_1 + c_2 f_2$$

one requires this to hold for $f(x) = x^2$, x and 1 separately, yielding the three simultaneous equations for c_0, c_1 and c_2:

$$\begin{aligned} 2x_0 &= c_0(x_0)^2 + c_1(x_0 + h)^2 + c_2(x_0 + 2h)^2 \\ 1 &= c_0(x_0) + c_1(x_0 + h) + c_2(x_0 + 2h) \\ 0 &= c_0(1) + c_1(1) + c_2(1) \end{aligned}$$

Solving for the c's, first simplifying by letting $x_0 = 0$ without loss of generality, and substituting back gives:

$$f_0' \approx \frac{-3f_0 + 4f_1 - f_2}{2h}$$

which agrees with the formula derivable using Section 3.1.4.

So far, only function values to the right of x_0 have been used in the method. One would expect this to give less accurate formulae than the use of function values on both sides of x_0, using closer and hence more pertinent information to the function near x_0. Indeed, beginning with

$$f_0' = c_{-1} f_{-1} + c_0 f_0 + c_1 f_1$$

yields the central difference formula

$$f_0' \approx \frac{f_1 - f_{-1}}{2h}$$

The technique can easily be applied to higher derivatives also. For instance, the assumption

$$f_0'' = c_{-1} f_{-1} + c_0 f_0 + c_1 f_1$$

gives the formula

$$f_0'' \approx \frac{f_1 - 2f_0 + f_{-1}}{h^2}$$

3.2. FUNCTION DERIVATIVES IN IRREGULAR GRID

3.2.1. Intuition

In the case of unequally spaced base-points x_{-1}, x_0 and x_1 where $x_1 - x_0 = s_E \Delta x$ and $x_0 - x_{-1} = s_W \Delta x$ $(0 \leq s_E,\ s_W \leq 1)$, consider the typical

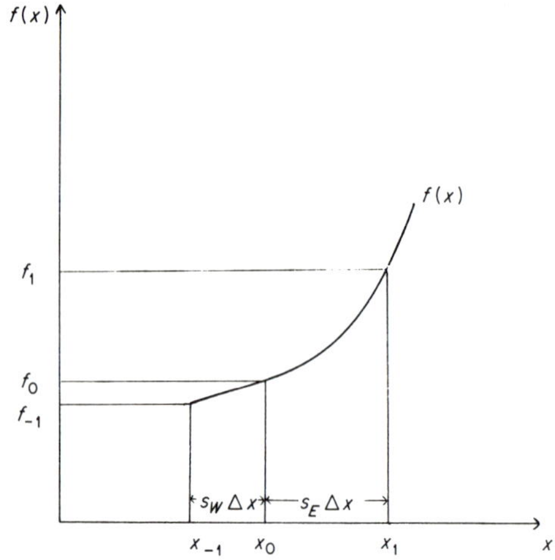

FIG. 3.3. Typical $f(x)$ showing function values f_{-1}, f_0 and f_1 taken at unequally spaced base-points x_{-1}, x_0 and x_1 where $x_1 - x_0 = s_E \Delta x$ and $x_0 - x_{-1} = s_W \Delta x$ $(0 \leq s_E, s_W \leq 1)$.

function shown in Fig. 3.3. Again one can make backward- and forward-difference estimates of $f'(x)$ at $x = x_0$. The forward one is estimated at $x_0 + \frac{1}{2}s_E\Delta x$ (the midpoint between x_0 and x_1) in terms of f_1 and f_0 as

$$\left.\frac{df}{dx}\right|_{x = x_0 + 1/2\, s_E \Delta x} = \frac{f_1 - f_0}{s_E \Delta x}$$

and the backward one at $x_0 - \frac{1}{2}s_W\Delta x$ (the midpoint between x_{-1} and x_0) in terms of f_{-1} and f_0 as

$$\left.\frac{df}{dx}\right|_{x = x_0 - 1/2\, s_W \Delta x} = \frac{f_0 - f_{-1}}{s_W \Delta x}$$

A central difference estimate simply as

$$\left.\frac{df}{dx}\right|_{x = x_0} = \frac{f_1 - f_{-1}}{(s_E + s_W)\,\Delta x}$$

is actually a best estimate at $x = x_{-1} + \frac{1}{2}(s_E + s_W)\,\Delta x = x_0 + \frac{1}{2}(s_E - s_W)\,\Delta x$ which is to the right of x_0 in the figure since $s_E > s_W$. It is better to take a weighted average of the two one-sided derivatives, weighted in favour of the nearer estimate, as

$$\left.\frac{\mathrm{d}f}{\mathrm{d}x}\right|_{x=x_0} = \frac{s_E}{s_E+s_W}\left.\frac{\mathrm{d}f}{\mathrm{d}x}\right|_{x=x_0-1/2\,s_W\Delta x} + \frac{s_W}{s_E+s_W}\left.\frac{\mathrm{d}f}{\mathrm{d}x}\right|_{x=x_0+1/2\,s_E\Delta x}$$

$$= \frac{s_E}{s_E+s_W}\cdot\frac{f_0-f_{-1}}{s_W\Delta x} + \frac{s_W}{s_E+s_W}\cdot\frac{f_1-f_0}{s_E\Delta x}$$

$$= \frac{1}{\Delta x}\left[\frac{s_W}{s_E(s_E+s_W)}f_1 - \frac{s_W-s_E}{s_E s_W}f_0 - \frac{s_E}{s_W(s_E+s_W)}f_{-1}\right]$$

Similarly, to estimate $f''(x)$ at $x = x_0$ the difference in the two one-sided derivatives, estimated at the midpoints of x_{-1} to x_0 and x_0 to x_1 respectively, is divided by the distance between them, $\frac{1}{2}(s_E + s_W)\,\Delta x$. This actually estimates $f''(x)$ at $x = x_0 + \frac{1}{2}(s_E - s_W)\,\Delta x$, which is slightly to the right of x_0 in the figure, but this value is retained as it is the best estimate for the value at x_0 based on the three function values. Thus

$$\left.\frac{\mathrm{d}^2f}{\mathrm{d}x^2}\right|_{x=x_0} \simeq \frac{\left(\left.\frac{\mathrm{d}f}{\mathrm{d}x}\right|_{x=x_0+1/2\,s_E\Delta x} - \left.\frac{\mathrm{d}f}{\mathrm{d}x}\right|_{x=x_0-1/2\,s_W\Delta x}\right)}{\left(\frac{(s_E+s_W)\,\Delta x}{2}\right)}$$

$$= \frac{\left(\frac{f_1-f_0}{s_E\Delta x} - \frac{f_0-f_{-1}}{s_W\Delta x}\right)}{\left(\frac{(s_E+s_W)\,\Delta x}{2}\right)}$$

$$= \frac{2}{(\Delta x)^2}\left[\frac{1}{s_E(s_E+s_W)}f_1 - \frac{1}{s_E s_W}f_0 + \frac{1}{s_W(s_E+s_W)}f_{-1}\right]$$

The reader is referred to Ames (1969) and Young (1962) for further details about this and subsequent sections.

3.2.2 Taylor Series

In the case of unequally spaced base-points as in Fig. 3.3, one proceeds in the same way to write out Taylor series expansions for $f(x_1) = f(x_0 + s_E\Delta x)$ and $f(x_{-1}) = f(x_0 - s_W\Delta x)$, thus obtaining

$$f_1 = f_0 + s_E\Delta x\,f_0' + \frac{(s_E\Delta x)^2}{2!}f_0'' + \frac{(s_E\Delta x)^3}{3!}f_0''' + \frac{(s_E\Delta x)^4}{4!}f_{0,1}^{\mathrm{iv}}$$

and

$$f_{-1} = f_0 - s_W\Delta x\,f_0' + \frac{(s_W\Delta x)^2}{2!}f_0'' - \frac{(s_W\Delta x)^3}{3!}f_0''' + \frac{(s_W\Delta x)^4}{4!}f_{-1,0}^{\mathrm{iv}}$$

There are several ways of manipulating these to get a formula for f_0' in terms of f_{-1}, f_0 and f_1. The most accurate is to truncate the series after the term in $(\Delta x)^2$, multiply the first by s_W^2 and the second by s_E^2, and subtraction then eliminates the $(\Delta x)^2$ term and yields the higher-order accuracy formula of order $(\Delta x)^2$ for f_0' as

$$\left.\frac{\mathrm{d}f}{\mathrm{d}x}\right|_{x=x_0} = \frac{1}{\Delta x}\left[\frac{s_W}{s_E(s_E+s_W)}f_1 - \frac{s_W - s_E}{s_E s_W}f_0 - \frac{s_E}{s_W(s_E+s_W)}f_{-1}\right] - \frac{s_E s_W(\Delta x)^2}{3!\,(s_E+s_W)}[s_E f'''_{0,1} + s_W f'''_{-1,0}]$$

Surprisingly, perhaps, the formula reads the same as the intuitive one with the addition of the error term. To obtain a $f''(x_0)$ formula the f_0' terms must be eliminated by multiplying the first by s_W and the second by s_E. Subtraction then yields

$$\left.\frac{\mathrm{d}^2f}{\mathrm{d}x^2}\right|_{x=x_0} = \frac{2}{(\Delta x)^2}\left[\frac{1}{s_E(s_E+s_W)}f_1 - \frac{1}{s_E s_W}f_0 + \frac{1}{s_W(s_E+s_W)}f_{-1}\right] + \frac{2\Delta x}{3!\,(s_E+s_W)}[s_E^2 f'''_{0,1} - s_W^2 f'''_{-1,0}]$$

which is the same as the intuitive formula with the addition of an error term of order Δx, one degree less than the corresponding formula for equally spaced base-points. The equally spaced base-point formula can of course be retrieved by setting $s_E = s_W = 1$, in which case cancellation of error terms yields a formula of order $(\Delta x)^2$.

3.2.3. Method of Undetermined Coefficients

All the formulae for unequally spaced base-points can also be derived using the method of undetermined coefficients. The technique can also cater for the use of more points, though the arithmetic gets heavy unless the more advanced method using Chebyshev polynomials is used (see Carnahan *et al.*, 1969; Gerald, 1970).

3.3. FDEs FOR STEADY-STATE PROBLEMS VIA PDE REPLACEMENT METHOD

A function of more than one variable possesses partial derivatives with respect to each of the independent variables. We shall be concerned principally with first- and second-order partial derivatives and will not need the cross-derivatives. The works of Ames (1969), Richtmyer & Morton (1967) and Smith (1965) provide illustrative reading. Thus if $T = T(x, y)$ is a function of two variables x and y, we shall be concerned with approximating the two first-order partial derivatives

$$\left.\frac{\partial T}{\partial x}\right|_{(x,y)=(x_0,y_0)} \quad \text{and} \quad \left.\frac{\partial T}{\partial y}\right|_{(x,y)=(x_0,y_0)}$$

and the two second-order partial derivatives

$$\left.\frac{\partial^2 T}{\partial x^2}\right|_{(x,y)=(x_0,y_0)} \quad \text{and} \quad \left.\frac{\partial^2 T}{\partial y^2}\right|_{(x,y)=(x_0,y_0)}$$

Since those with respect to x are simply the ordinary derivatives of T with respect to x whilst keeping y constant, and vice versa, it follows that all the derivative ideas and formulae of Sections 3.1 and 3.2 are directly applicable. One has, of course, to use points on a line of constant $y = y_0$ when obtaining from Sections 3.1 and 3.2 a formula for the partial x-derivatives, and vice versa.

3.3.1. Cartesian Coordinates: Uniform Grid

The discussion is temporarily restricted to a 3-D conducting region with uniform conductivity k in cartesian coordinates with a uniform grid system arranged throughout the domain in which solution is sought. For now, only internal points will be considered. If all boundaries lie on surfaces with one coordinate equal to a constant $x = a$, for example, and the grid system is arranged to touch the boundaries directly, then the internal temperature field problem can be solved if boundary temperatures are known explicitly (Dirichlet conditions). Figure 3.4 shows a typical grid point P with its immediate six neighbours, labelled N (north), S (south), E (east), W (west), I (in), O (out). Let $\Delta = \Delta x = \Delta y = \Delta z$ be the uniform grid spacing. Clearly, then, the governing steady-state equation (eqn. (2.2.4) (i) of Chapter 2 with $\partial T/\partial t = 0$):

$$\frac{\partial^2 T}{\partial x^2} + \frac{\partial^2 T}{\partial y^2} + \frac{\partial^2 T}{\partial z^2} + \frac{H}{k} = 0$$

may be approximated, using the results of Section 3.1, by

$$\frac{T_O - 2T_P + T_I}{\Delta^2} + \frac{T_E - 2T_P + T_W}{\Delta^2} + \frac{T_N - 2T_P + T_S}{\Delta^2} + \frac{H}{k} = 0$$

where, if H varies with position or temperature, it is evaluated at the point P.

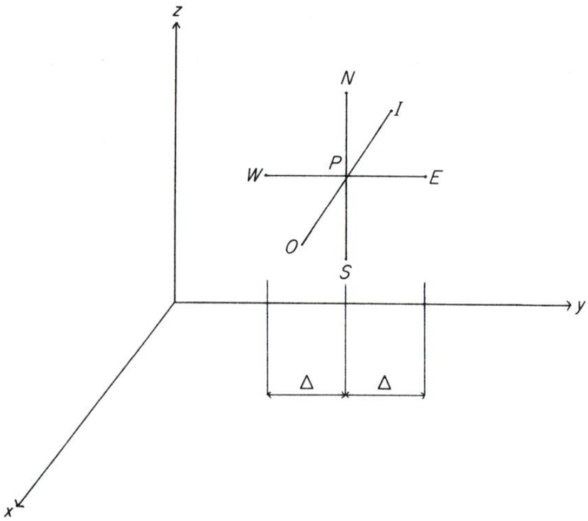

FIG. 3.4. Typical point P with its six neighbours in a 3-D cartesian coordinate region with uniform grid spacing.

This expression simplifies to

$$T_O + T_I + T_E + T_W + T_N + T_S - 6T_P + \frac{H}{k}\Delta^2 = 0 \qquad (3.3.1)$$

Clearly, for 2-D (x, y) conduction regions the analogous expression is

$$T_O + T_I + T_E + T_W - 4T_P + \frac{H}{k}\Delta^2 = 0 \qquad (3.3.2)$$

and for 1-D (x) heat flows:

$$T_O + T_I - 2T_P + \frac{H}{k}\Delta^2 = 0 \qquad (3.3.3)$$

Obviously with no internal heat generation, the last term on the left-hand side in each of these three expressions is zero. With $H = 0$, eqn. (3.3.2) is generally referred to as the five-point formula, and represented pictorially as

$$\begin{bmatrix} & 1 & \\ 1 & -4 & 1 \\ & 1 & \end{bmatrix} T_P = 0$$

where the error is $0(\Delta^4)$. How to handle time-dependent problems is discussed in Section 3.5.

3.3.2. Cartesian Coordinates: Irregular Grid

If now the uniform mesh restriction is removed and grid spacing from the point P to the points N, S, E, W, I and O is represented by $s_N\Delta, s_S\Delta, \ldots, s_O\Delta$, the formulae of Section 3.2 can be used. For the analogous 3-D problem with non-uniform grid to the one just considered with a uniform grid, the steady-state equation may be represented as

$$\frac{2}{\Delta^2}\left[\frac{1}{s_N(s_N+s_S)}T_N - \frac{1}{s_N s_S}T_P + \frac{1}{s_S(s_N+s_S)}T_S\right]$$
$$+\frac{2}{\Delta^2}\left[\frac{1}{s_E(s_E+s_W)}T_E - \frac{1}{s_E s_W}T_P + \frac{1}{s_W(s_E+s_W)}T_W\right]$$
$$+\frac{2}{\Delta^2}\left[\frac{1}{s_O(s_I+s_O)}T_O - \frac{1}{s_I s_O}T_P + \frac{1}{s_I(s_I+s_O)}T_I\right] + \frac{H}{k} = 0 \quad (3.3.4)$$

where H is evaluated at P if it varies with position or temperature. Note that setting all the s's equal to unity retrieves the uniform grid formula. Degenerate versions from this equation are easily derivable for 2-D and 1-D problems, with the obvious neglect of the last term when internal heat generation is lacking. Unfortunately the equations do not simplify any further, but in Section 3.4 it will be seen that the same formula may be derived from energy balance considerations, and coefficients can easily be stored in a computer program in locations a_N, a_S, a_E, a_W, a_I, a_O, a_P and a_C defined by the equivalence of

$$a_N T_N + a_S T_S + a_E T_E + a_W T_W + a_I T_I + a_O T_O + a_P T_P + a_C = 0 \quad (3.3.5)$$

to the above equation, eqn. (3.3.4), or the uniform grid version, eqn. (3.3.1). For example, in a uniform grid:

$$a_N = a_S = a_E = a_W = a_I = a_O = 1$$
$$a_P = -6$$
$$a_C = \frac{H}{k}\Delta^2$$

and in a non-uniform grid:

$$a_N = \frac{2}{\Delta^2}\frac{1}{s_N(s_N+s_S)} \quad \text{and similarly for } a_S,\ a_E,\ a_W,\ a_I \text{ and } a_O$$

$$a_P = -\frac{2}{\Delta^2}\left[\frac{1}{s_N s_S} + \frac{1}{s_E s_W} + \frac{1}{s_I s_O}\right]$$

$$a_C = \frac{H}{k}$$

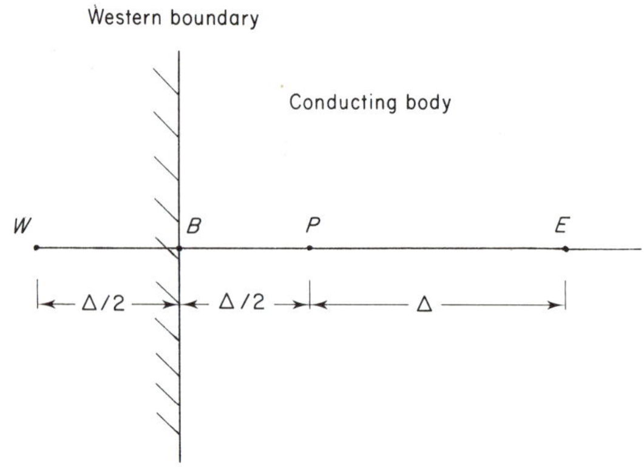

FIG. 3.5. 2-D steady-state problem with western boundary lying midway between.

Later, in Chapters 5 *et seq*., an equation of this type will be used for updating T_P in a process iterating towards the solution at P of the heat conduction problem.

Rather than arranging for the grid lines to lie directly on the boundaries, it is sometimes convenient, when some boundaries have convective (Neumann) conditions to apply, to have the boundaries lying midway between two grid lines. This is illustrated for a 2-D steady-state problem in Fig. 3.5 for the case of a western boundary of known temperature T_B. The point W lies outside the solution domain and is in fact not used. The more complicated case in which T_B is not known, but is implied via the application of a convective (Neumann, derivative) boundary condition, is dealt with in Section 4.3 of Chapter 4. The FDE for the point P is written down in terms of values at the points E and B, using the irregular grid derivative formulation of Section 3.2 with $s_E = 1$ and $s_W = \frac{1}{2}$. Thus the $\partial^2 T/\partial x^2$ term is approximated by

$$\frac{2}{\Delta^2}\left[\frac{1}{(1)(1+\frac{1}{2})}T_E - \frac{1}{(1)\cdot(\frac{1}{2})}T_P + \frac{1}{(\frac{1}{2})(1+\frac{1}{2})}T_B\right]$$

which may be simplified to the form

$$\frac{2}{\Delta^2}\left[\frac{2}{3}T_E - 2\,T_P + \frac{4}{3}T_B\right]$$

instead of the previously given formulation. If the problem is 1-D, the appropriate FDE would be

$$T_E - 3T_P + 2T_B = 0 \tag{3.3.6}$$

provided $H = 0$ and steady state is assumed. Alternatively, but in an analogous technique to the one used with convective boundary conditions, the usual formula could be applied at P:

$$\frac{1}{\Delta^2}[T_E - 2T_P + T_W] = 0$$

together with the supplementary equation which linearly extrapolates to estimate the value at W:

$$T_W = T_B + (T_B - T_P)$$

Elimination of T_W from these two equations yields a similar formula for T_P which is, however, only of first-order accuracy. Of course, the three points E, P and B could be used to extrapolate for the value at W. However, these methods, which require an update after each iteration over the internal field values, are not generally recommended. They are useful in cases where a simple formula is to be applied over many internal points and it is good not to have to store the coupling coefficients explicitly in location $a_N, a_S, \ldots$ for each point P. It is convenient then to apply the same formula to the near boundary points as well, but this must be supplemented by the updating of the exterior fictitious points, like the W point considered above when the western boundary is handled like this.

If there is a further complication, like an irregular grid combined with boundaries lying midway between adjacent grid lines, a formula for $\partial^2 T/\partial x^2$ at P based on values at E, P and B can easily be obtained using Section 3.2. The method of undetermined coefficients is also useful in these and more complicated situations. Again, the energy balance method derives precisely the same formulae when applied over E, P and B, for example.

3.3.3. More Accurate Formulae

The discussion is now restricted to 2-D steady-state conduction problems with zero internal heat generation using a uniform square grid system with spacing Δ. In this case the usual FDE is represented pictorially as

$$\begin{bmatrix} & 1 & \\ 1 & -4 & 1 \\ & 1 & \end{bmatrix} T_P = 0$$

It can also be shown (Thom & Apelt, 1961) that the pictorial formula

$$\begin{bmatrix} 1 & & 1 \\ & -4 & \\ 1 & & 1 \end{bmatrix} T_P = 0 \qquad (3.3.7)$$

is also an approximation to the same governing PDE. A more accurate FDE can be obtained by combining these two, the best combination being the nine-point formula represented pictorially by

$$\begin{bmatrix} 1 & 4 & 1 \\ 4 & -20 & 4 \\ 1 & 4 & 1 \end{bmatrix} T_P = 0 \tag{3.3.8}$$

where the error is $0(\Delta^6)$.

Though this and higher-order accuracy formulae look attractive, they present more complexity in updating T_P in an iterative solution process and affect the rate of convergence (or divergence) of the solution process.

3.4. FDEs FOR STEADY-STATE PROBLEMS VIA ENERGY BALANCE CONSIDERATION METHODS

The appropriate FDE for the temperature at a node point can also be derived by energy balance considerations and the formulae of Chapter 2, Section 2.1. Each node point is taken as representative of a region, called a cell control volume, in the system. Heat flows across the surfaces of such a region from adjoining regions are considered and a formula derived for the node temperature in terms of temperatures at neighbouring node points. The technique leads to the same results as were obtained in Section 3.3, with which there is complete agreement. Furthermore it is often convenient and easy to apply, especially to variable grids, convective boundary conditions, odd-shaped regions, etc.

3.4.1. Cartesian Coordinates: Uniform Grid

Consider for clarity the cell control volume around the point P in a 3-D cartesian system with uniform grid spacing $\Delta = \Delta x = \Delta y = \Delta z$, as shown in Fig. 3.6 together with its six immediate neighbours. All the cell faces (shown shaded for node E) have area Δ^2 and the distance from any of the six neighbours to P is Δ. Provided P is not a boundary point, and assuming that if W lies on the western boundary then T_W, for example, is known, all internal points P can be considered as follows. The thermal conductances K_{ij} between a node i and its adjacent node j are calculated and then the rate of heat flow Q_{ij} from j to i is

$$Q_{ij} = K_{ij}(T_j - T_i) \tag{3.4.1}$$

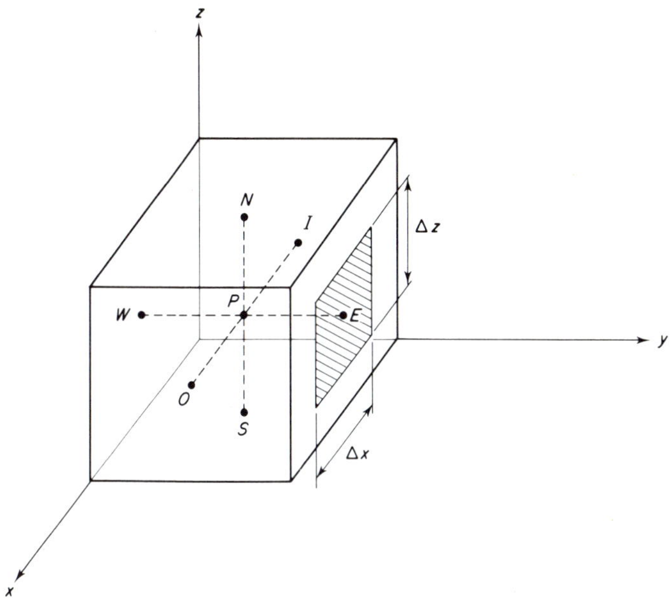

FIG. 3.6. The cell control volume around a typical point P in a 3-D cartesian coordinate region with uniform grid spacing, together with its six neighbours.

Under steady-state conditions, the energy balance for the node i gives the following equation:

$$\sum_j Q_{ij} + H \cdot V_i = \sum_j K_{ij}(T_j - T_i) + H \cdot V_i = 0 \qquad (3.4.2)$$

(where the summation is over all the cell faces), using both the heat flowing into the cell control volume V_i through the six faces plus the internal heat generation H (W/m^3). The cell control volume V_i for node i is

$$\begin{aligned} V_i &= \Delta x \Delta y \Delta z \\ &= \Delta^3 \end{aligned}$$

since the grid spacing is all uniform and equal to Δ. Since the thermal conductances for internal node points are calculated using the form

$$K_{ij} = \frac{kA_{ij}}{L_{ij}} \qquad (3.4.3)$$

eqn. (3.4.1) is analogous to eqn. (2.1.2) of Chapter 2. Here k is the thermal conductivity, A_{ij} is the average surface area perpendicular to the direction of heat flow between j and i, and L_{ij} is the distance between the two nodes i and j. At boundary points i

$$K_{ij} = h_j A_{ij} \tag{3.4.4}$$

where h_j is the average surface heat transfer coefficient on the surface of area A_{ij} exposed to the environment at temperature T_j. Then if, for example, P is a near-boundary point whose cell control volume western face of area A_w lies on the boundary exposed to an ambient temperature T_A through the surface heat transfer coefficient h, then the rate of heat flow from the ambient to the point P is

$$Q_{ij} = K_{ij}(T_j - T_i)$$

or

$$Q_{PA} = hA_w(T_A - T_P) \tag{3.4.5}$$

It will be seen in Chapter 4, Section 4.3, that the incorporation of this and similar boundary heat flow considerations yields the same formulae as the PDE replacement method, with elimination of the temperature of the immediate exterior fictitious point.

Returning to the 3-D cartesian problem and considering the energy balance for the point P with its six neighbours shown in Fig. 3.6, one obtains, using eqns. (3.4.1) and (3.4.3):

$$k\frac{\Delta^2}{\Delta}(T_N - T_P) + k\frac{\Delta^2}{\Delta}(T_S - T_P) + k\frac{\Delta^2}{\Delta}(T_E - T_P) + k\frac{\Delta^2}{\Delta}(T_W - T_P)$$
$$+ k\frac{\Delta^2}{\Delta}(T_I - T_P) + k\frac{\Delta^2}{\Delta}(T_O - T_P) + H \cdot \Delta^3 = 0$$

or

$$\frac{k}{\Delta^2}(T_N - 2T_P + T_S) + \frac{k}{\Delta^2}(T_E - 2T_P + T_W) + \frac{k}{\Delta^2}(T_I - 2T_P + T_O) + H = 0$$

which is now seen to be equivalent to direct PDE replacement. Continuing, this is similarly equal to the form

$$T_N + T_S + T_E + T_W + T_I + T_O - 6T_P + \frac{H \cdot \Delta^2}{k} = 0$$

as before (eqn. (3.3.1)). As in Section 3.3, 2-D and 1-D versions are easily obtained in a similar fashion using a cell control volume with arbitrary size unity in the irrelevant distances. For example, consider the point P of a 2-D region, shown in Fig. 3.7 with its four neighbours. Considering heat flows

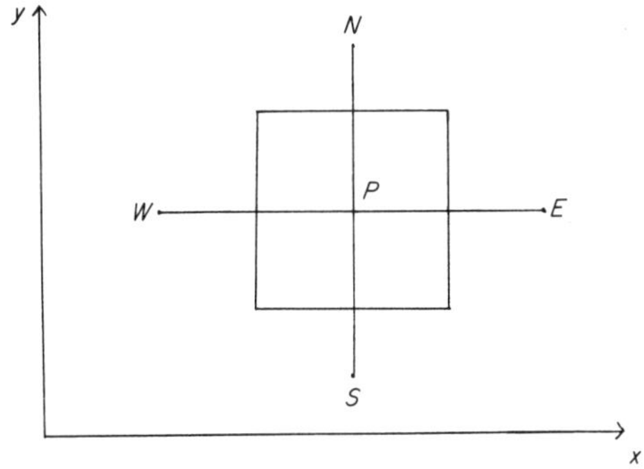

FIG. 3.7. The cell control volume around a typical point P in a 2-D cartesian coordinate region with uniform grid spacing, together with its four neighbours.

through the four faces:

$$k\frac{\Delta\cdot(1)}{\Delta}(T_N - T_P) + k\frac{\Delta\cdot(1)}{\Delta}(T_S - T_P) + k\frac{\Delta\cdot(1)}{\Delta}(T_E - T_P) + k\frac{\Delta\cdot(1)}{\Delta}(T_W - T_P) + H\cdot\Delta^2\cdot(1) = 0$$

from which the familiar (and equivalent to eqn. (3.3.2))

$$T_N + T_S + T_E + T_W - 4T_P + \frac{H}{k}\Delta^2 = 0$$

results. Of course, with no internal heat generation the last term in the above equations with H is set to zero.

3.4.2. Cartesian Coordinates: Irregular Grid

The energy balance technique works just as easily for non-uniform problems. As an example, consider a 3-D problem similar to the one portrayed in Fig. 3.6, except that variable grid spacing is now allowed. The spacings from the point P to N, S, E, W, I and O are now represented by $s_N\Delta$, $s_S\Delta$, . . ., $s_O\Delta$ and the following cell control volume measurements are noted:

$$\text{volume } V = \frac{1}{2}(s_N + s_S)\,\Delta\cdot\frac{1}{2}(s_E + s_W)\,\Delta\cdot\frac{1}{2}(s_I + s_O)\,\Delta$$

$$= \frac{\Delta^3}{8}(s_N + s_S)(s_E + s_W)(s_I + s_O)$$

The north and south cell areas (A_n and A_s) are both given by

$$A_n = A_s = \frac{1}{2}(s_E + s_W)\,\Delta \cdot \frac{1}{2}(s_I + s_O)\,\Delta$$

$$= \frac{\Delta^2}{4}(s_E + s_W)(s_I + s_O)$$

$$= \frac{V}{\left[\dfrac{\Delta}{2}(s_N + s_S)\right]}$$

Similar expressions can be derived for the other nodes.

Consider now the heat balance on this cell control volume:

$$k\frac{V}{\dfrac{\Delta}{2}(s_N + s_S)} \cdot \frac{1}{s_N \Delta}(T_N - T_P) + k\frac{V}{\dfrac{\Delta}{2}(s_N + s_S)} \cdot \frac{1}{s_S \Delta}(T_S - T_P)$$

+ four similar terms for E, W, I and O contributions
$+ H \cdot V = 0$

where V is as given above. This is equivalent to

$$\frac{2}{\Delta^2}\left[\frac{1}{s_N(s_N + s_S)}T_N - \frac{1}{s_N s_S}T_P + \frac{1}{s_S(s_N + s_S)}T_S\right]$$

+ two similar terms

$$+ \frac{H}{k} = 0$$

which is precisely the same as eqn. (3.3.3) which was obtained by direct PDE replacement methods. The middle term of the first bracket follows since

$$\frac{1}{s_N(s_N + s_S)} + \frac{1}{s_S(s_N + s_S)} = \frac{s_S + s_N}{s_N s_S (s_N + s_S)} = \frac{1}{s_N s_S}$$

A similar discussion shows the equivalence of the energy balance technique with the PDE replacement technique for irregular 2-D and 1-D problems. Hence either technique which is most convenient for the problem in hand is recommended.

The equation for T_P, when T_E and T_B are known, in Fig. 3.5 is found very simply by this method. One considers a small cell control volume around P whose boundaries bisect BP and PE. Since a 1-D problem is assumed, take unity to be the cell dimensions in the y- and z-directions. Considering now the energy balance, one has

$$k\frac{1^2}{\Delta}(T_E - T_P) + k\frac{1^2}{\Delta/2}(T_B - T_P) = 0$$

from which

$$\frac{k}{\Delta}(T_E - T_P + 2T_B - 2T_P) = 0$$

or, equivalently,

$$T_E - 3T_P + 2T_B = 0$$

which is precisely eqn. (3.3.6), obtained by the PDE replacement method.

3.4.3. Summary

Two important methods of deriving steady-state FDEs have now been described: the energy balance method and the direct replacement of PDEs method. The methods are discussed again with illustrative examples in Chapter 5 and given further application in the problems discussed in Chapter 8.

3.5. FDEs FOR TRANSIENT PROBLEMS

Transient or unsteady-state heat transfer problems are characterised by time-dependent heat fluxes and temperature fields. FDEs may be obtained via PDE replacement or energy balance techniques, as is now illustrated for uniform spatial grid systems. Obvious modifications to cater for irregular grids using the ideas of Sections 3.3 or 3.4 may also be incorporated.

3.5.1. PDE Replacement

Consider first a 3-D conducting region with uniform conductivity k in cartesian coordinates with a uniform grid system arranged throughout the domain in which solution is sought. Take $\Delta = \Delta x = \Delta y = \Delta z$ and concentrate on a typical internal point like P of Fig. 3.4 where its six immediate neighbours are also shown. For simplicity, let us assume no internal heat generation as well. The governing PDE is then

$$\frac{\partial^2 T}{\partial x^2} + \frac{\partial^2 T}{\partial y^2} + \frac{\partial^2 T}{\partial z^2} = \frac{1}{\alpha}\frac{\partial T}{\partial t} \tag{3.5.1}$$

where the thermal diffusivity $\alpha = k/\rho C_p$. In an explicit formulation of this equation, the left-hand-side terms are evaluated at time t, and a forward-difference representation is used for the right-hand-side term which involves $T(t + \Delta t)$ and $T(t)$ at the point P only. Using the results of Section 3.1, the FDE for the point P is

$$\frac{T_O - 2T_P + T_I}{\Delta^2} + \frac{T_E - 2T_P + T_W}{\Delta^2} + \frac{T_N - 2T_P + T_S}{\Delta^2} = \frac{1}{\alpha} \cdot \left(\frac{T'_P - T_P}{\Delta t} \right) \tag{3.5.2}$$

with the notation $T'_P = T_P(t + \Delta t)$

$$T_P = T_P(t)$$

and all other non-primed quantities are evaluated at time-level t. This expression reduces to the single form

$$T_O + T_I + T_E + T_W + T_N + T_S - 6T_P = \frac{1}{\text{Fo}}(T'_P - T_P)$$

or

$$T_O + T_I + T_E + T_W + T_N + T_S - \left(6 - \frac{1}{\text{Fo}}\right) T_P = \frac{1}{\text{Fo}} T'_P \tag{3.5.3}$$

where the Fourier number $\text{Fo} = \alpha \Delta t/\Delta^2 = k\Delta t/\rho C_p \Delta^2$. This expression allows $T'_P = T_P(t + \Delta t)$ to be calculated simply in terms of its own and its six neighbouring temperatures at time-level t, which are all known. Similar derivation for a 2-D (x, y) conducting region yields the analogous expression

$$T_O + T_I + T_E + T_W - \left(4 - \frac{1}{\text{Fo}}\right) T_P = \frac{1}{\text{Fo}} T'_P \tag{3.5.4}$$

and for 1-D (x) heat flows:

$$T_O + T_I - \left(2 - \frac{1}{\text{Fo}}\right) T_P = \frac{1}{\text{Fo}} T'_P \tag{3.5.5}$$

There are considerations with regard to stability, accuracy and convergence of these formulae which limit the maximum value of Δt for a given Δ spatial subdivision, and which will be discussed in Chapter 4, Section 4.5. The above three equations may be supplemented if there is internal heat generation by the addition of $(H/k)\Delta^2$ evaluated at P at the previous time-level t.

A similar formulation can lead to what is known as an implicit equation for T'_P, the equation containing other temperatures at nearby points at the new time-level $t + \Delta t$. Hence T'_P cannot be simply computed from previous known values at time-level t and one has to solve for each time-step the set of simultaneous equations for the new temperatures together; the number of such equations is equal to the total number of nodes at which T is unknown. The FDE for the point P, evaluating all the terms at time-level $t + \Delta t$, is

$$\frac{T'_O - 2T'_P + T'_I}{\Delta^2} + \frac{T'_E - 2T'_P + T'_W}{\Delta^2} + \frac{T'_N - 2T'_P + T'_S}{\Delta^2} = \frac{1}{\alpha} \cdot \left(\frac{T'_P - T_P}{\Delta t}\right) \tag{3.5.6}$$

with the same notation as before. This simplifies to

$$T'_O + T'_I + T'_E + T'_W + T'_N + T'_S - \left(6 + \frac{1}{\text{Fo}}\right) T'_P = -\frac{1}{\text{Fo}} T_P \tag{3.5.7}$$

The 2-D (x, y) version is

$$T'_O + T'_I + T'_E + T'_W - \left(4 + \frac{1}{\text{Fo}}\right) T'_P = -\frac{1}{\text{Fo}} T_P \tag{3.5.8}$$

and for 1-D (x) heat flows:

$$T'_O + T'_I - \left(2 + \frac{1}{\text{Fo}}\right) T'_P = -\frac{1}{\text{Fo}} T_P \tag{3.5.9}$$

These three formulae are stable with regard to the propagation of rounding error regardless of the Δt time-step used; however, they are less accurate and the work-load required to advance one time-step is considerably more than the simpler explicit formulation, except for the case of a 1-D transient problem where, as will be seen later in Chapter 7, Section 7.5, a simple tridiagonal matrix algorithm (TDMA) for solution is readily applicable. The above three formulae are so-called fully implicit. They are a special case of a generalised formulation which evaluates the spatial derivatives at some time-location between t and $t + \Delta t$, obtained by taking a weighted average of the explicit and fully implicit formulae for the spatial derivatives. If the weighting parameter λ $(0 \leq \lambda \leq 1)$ is used via

$$\text{Value taken} = \lambda\ (\text{value at } t + \Delta t) + (1 - \lambda)\ (\text{value at } t) \tag{3.5.10}$$

then the 3-D problem is represented by

$$\lambda(T'_O + T'_I + T'_E + T'_W + T'_N + T'_S - 6T'_P) + (1 - \lambda)(T_O + T_I + T_E + T_W + T_N + T_S - 6T_P) = \frac{1}{\text{Fo}}(T'_P - T_P) \tag{3.5.11}$$

from which

$$\lambda(T'_O + T'_I + T'_E + T'_W + T'_N + T'_S) - \left(6\lambda + \frac{1}{\text{Fo}}\right)T'_P$$
$$= -(1-\lambda)(T_O + T_I + T_E + T_W + T_N + T_S)$$
$$+ \left[6(1-\lambda) - \frac{1}{\text{Fo}}\right]T_P \quad (3.5.12)$$

Setting $\lambda = 0$, 1, 0·5 retrieves the explicit, fully implicit and Crank–Nicolson mid-difference schemes respectively, and there are reasons why the last of these is preferable to the others (see Chapter 4, Section 4.5). The 2-D (x, y) and 1-D (x) analogous expressions are

$$\lambda(T'_O + T'_I + T'_E + T'_W) - \left(4\lambda + \frac{1}{\text{Fo}}\right)T'_P$$
$$= -(1-\lambda)(T_O + T_I + T_E + T_W) + \left[4(1-\lambda) - \frac{1}{\text{Fo}}\right]T_P \quad (3.5.13)$$

and

$$\lambda(T'_O + T'_I) - \left(2\lambda + \frac{1}{\text{Fo}}\right)T'_P$$
$$= -(1-\lambda)(T_O + T_I) + \left[2(1-\lambda) - \frac{1}{\text{Fo}}\right]T_P \quad (3.5.14)$$

where the last of these is particularly useful, since the direct tridiagonal matrix algorithm (TDMA) is rapid for its solution, once the edge-boundary conditions have been incorporated.

3.5.2. Energy Balance Techniques

Exactly the same FDE for T'_P results from the use of energy balance techniques. Considering the same configuration as with the PDE replacement method (Fig. 3.4), one includes a control volume around P as shown in Fig. 3.6 and applies energy balance principles to this volume. The actual cell control volume associated with node P is equal to $(\Delta x\, \Delta y\, \Delta z)$. The conductances from each of the six neighbours I, O, N, S, E and W to P are calculated in terms of k, A_{ij} and L_{ij} as was done previously using the energy method for steady-state problems (see eqn. (3.4.3)). Recalling this equation:

$$K_{ij} = \frac{kA_{ij}}{L_{ij}} \quad (3.5.15)$$

where k is the thermal conductivity, A_{ij} is the average surface perpendicular to the direction of heat flow between points j and i (the area of one of the faces of the cube in Fig. 3.6) and L_{ij} is the distance between the two nodes i and j, consider the energy balance for the node P, which is surrounded by six adjacent nodes I, O, N, S, E and W, during a small time-interval Δt. During this time-interval one assumes that the temperatures at each of the nearby nodes remain constant and that heat fluxes calculated at a fixed time-level between t and $t + \Delta t$ will be operative throughout the small time-interval Δt. The heat flow into the node P (that is, its control volume) will cause a change $\Delta T_P = T'_P - T_P$ in its temperature, using the ideas described in Chapter 2, Section 2.2. The heat flows are given by eqns. (3.4.1) and (3.4.3):

$$Q_{ij} = \frac{kA_{ij}}{L_{ij}}(T_j - T_i) \tag{3.5.16}$$

in general, or a typical one for Fig. 3.6 is

$$Q_{PN} = k\frac{\Delta^2}{\Delta}(T_N - T_P) \tag{3.5.17}$$

Thus

$$\sum_{j = I, O, N, S, E \text{ and } W} k\frac{\Delta^2}{\Delta}(T_j - T_P) = \rho C_p V_P \frac{\Delta T_P}{\Delta t} \tag{3.5.18}$$

where ρ and C_p are density (kg/m^3) and specific heat (J/kg K) of the body, and $V_P = \Delta x\, \Delta y\, \Delta z = \Delta^3$ is the volume (m^3) of the cell control volume for P. If there is heat generation in the system, a term $H \cdot V_P$ must be added to the left-hand side of eqn. (3.5.18). This expression can be used in general for the nodal equation for any internal node P, but boundary nodes are handled similarly, using eqn. (3.4.4) for the surface thermal conductance K_{ij}; it is recalled here that

$$K_{ij} = h_j A_{ij} \tag{3.5.19}$$

where h_j is the average surface heat transfer coefficient on the surface of area A_{ij} exposed to the environment at temperature T_j. These details are considered in Chapter 4, Section 4.3. Different ways of taking $\Delta T_P/\Delta t$ in eqn. (3.5.18) lead to different forms of the FDE; but it is usually taken as a forward-difference from t to $t + \Delta t$:

$$\frac{\Delta T_P}{\Delta t} = \frac{T_P(t + \Delta t) - T_P(t)}{\Delta t} = \frac{T'_P - T_P}{\Delta t} \tag{3.5.20}$$

Likewise the choice of time-level at which the left-hand side is evaluated (t, $t + \Delta t$ or at some time-level between them given by

$$\lambda \text{ (Value at } t + \Delta t) + (1 - \lambda) \text{ (Value at } t)$$

where $0 \le \lambda \le 1$) yields the explicit, fully implicit and weighted average intermediate values of the spatial derivative terms. Putting $\lambda = 0{\cdot}5$ generates the Crank–Nicolson scheme. To show the equivalence of the energy balance and the PDE replacement techniques, consider the weighted-average energy balance formula for left-hand-side terms of eqn. (3.5.18) and the forward-difference formulation, eqn. (3.5.20), for the right-hand side. The result is

$$[\lambda k \Delta (T'_O - T'_P) + (1 - \lambda) k \Delta (T_O - T_P)]$$
$$+ \text{ five similar terms with } O \text{ replaced by } I, N, S, E \text{ and } W$$
$$= \rho C_p \Delta^3 \left(\frac{T'_P - T_P}{\Delta t} \right) \quad (3.5.21)$$

which rearranges to

$$\lambda (T'_O + T'_I + T'_N + T'_S + T'_E + T'_W - 6T'_P)$$
$$+ (1 - \lambda)(T_O + T_I + T_N + T_S + T_E + T_W - 6\,T_P)$$
$$= \frac{\rho C_p \Delta^2}{k\,\Delta t} (T'_P - T_P) \quad (3.5.22)$$

which is exactly eqn. (3.5.11) and hence eqn. (3.5.12) on substituting for the Fourier number $\text{Fo} = k\Delta t/\rho C_p \Delta^2$. Similar derivatives of the degenerate 2-D and 1-D versions give FDEs which are exactly the same as eqns. (3.5.13) and (3.5.14). Obviously, taking $\lambda = 0$ or $\lambda = 1$ yields equivalent formulae to eqns. (3.5.3)–(3.5.5) and eqns. (3.5.7)–(3.5.9) respectively. Incorporation of an internal heat generation term again yields equivalent FDEs regardless of whether PDE replacement or energy balance methods are used.

3.5.3. Summary

Two important methods of deriving FDEs for transient heat flow problems have been described: the energy balance method and the direct replacement of PDEs method. In the same way that spatial derivatives are replaced by increments of temperature and distance, so time derivatives are replaced by increments of temperature and time. The size of the time-step cannot be chosen arbitrarily for explicit FDEs because of stability considerations. There is more freedom of choice with implicit FDEs.

The methods are discussed again with illustrative examples in Chapter 6 and given further application in the problems discussed in Chapter 8.

BIBLIOGRAPHY

ABRAMOWITZ, M. & STEGUN, I. A. (eds.) (1964). *Handbook of Mathematical Functions*. Applied Mathematics Series No. 55, National Bureau of Standards, US Dept. of Commerce, Washington, DC.

AMES, W. F. (1969). *Numerical Methods for Partial Differential Equations*. Nelson, London.

CARNAHAN, B., LUTHER, H. A. & WILKES, J. O. (1969). *Applied Numerical Methods*. Wiley, New York.

DUSINBERRE, G. M. (1949). *Numerical Analysis of Heat Flow*. McGraw-Hill, New York.

DUSINBERRE, G. M. (1961). *Heat Transfer Calculations by Finite Differences*. International Textbook Co., Scranton, Pa.

GERALD, C. F. (1970). *Applied Numerical Analysis*. Addison-Wesley, Reading, Mass.

NATIONAL PHYSICAL LABORATORY (1961). *Modern Computing Methods*, 2nd ed. Notes on Applied Science Series No. 16, HMSO, London.

NAUTICAL ALMANAC OFFICE (1956). *Interpolation and Allied Tables*. HMSO, London.

RAZELOS, P. (1973). Methods of obtaining approximate solutions. Section 4, pp. 4–1 to 4–78, in *Handbook of Heat Transfer*, ed. W. M. Rohsenow and J. P. Hartnett. McGraw-Hill, New York.

RICHTMYER, R. D. & MORTON, K. W. (1967). *Difference Methods for Initial Value Problems*. Interscience, New York.

SCHUH, H. (1965). *Heat Transfer in Structures*. Pergamon Press, Oxford.

SMITH, G. D. (1965). *Numerical Solution of Partial Differential Equations*. Oxford University Press, London.

THOM, A. & APELT, C. J. (1961). *Field Computations in Engineering and Physics*. Van Nostrand, London.

YOUNG, D. (1962). The numerical solution of elliptic and parabolic partial differential equations. In: *A Survey of Numerical Analysis*, ed. J. Todd. McGraw-Hill, New York.

CHAPTER 4

Some Complications with Finite Difference Equations

4.1. FDEs IN NON-CARTESIAN COORDINATE SYSTEMS

It is easier to represent the shape of some objects by a coordinate system other than cartesian. The general equation of heat conduction has already been described in both cylindrical polar and spherical polar coordinates in Section 2.3 of Chapter 2. We shall now see that these equations can be represented in FDE form by using the same methods as those used for the cartesian systems.

4.1.1. Cylindrical Polar Coordinates

The general equation of heat conduction in these coordinates (taking the thermal conductivity as constant) is

$$\frac{\partial^2 T}{\partial r^2} + \frac{1}{r}\frac{\partial T}{\partial r} + \frac{1}{r^2}\frac{\partial^2 T}{\partial \theta^2} + \frac{\partial^2 T}{\partial z^2} + \frac{H}{k} = \frac{1}{\alpha}\frac{\partial T}{\partial t} \tag{4.1.1}$$

(The detailed derivation is given in Section 2.3 of Chapter 2.) The FDE which replaces the above PDE will be derived by the two main methods previously explained, direct replacement and energy balance.

4.1.2. Direct Replacement

The geometric configuration is shown in Figs. 4.1 and 4.2. Equal increments of z, r and θ are used. Directly replacing the above PDE will give the following FDE for node 0:

$$\left[\frac{T_2 - 2T_0 + T_4}{(\Delta r)^2} + \frac{T_2 - T_4}{2r_0 \Delta r} + \frac{T_1 - 2T_0 + T_3}{r_0^2(\Delta\theta)^2} + \frac{T_5 - 2T_0 + T_6}{(\Delta z)^2} + \frac{H}{k}\right] = \frac{(T_0' - T_0)}{\alpha\,\Delta t} \tag{4.1.2}$$

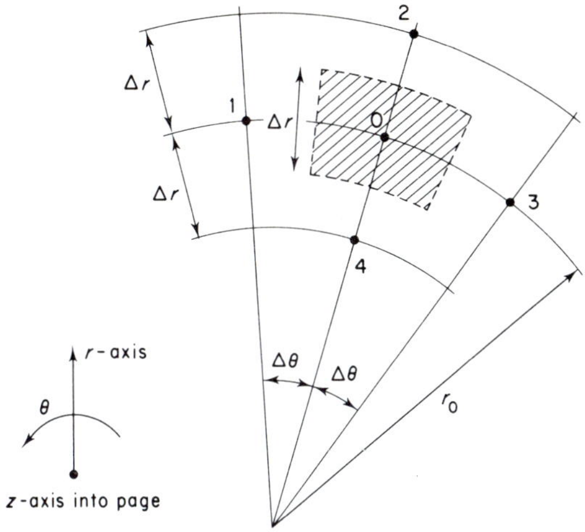

FIG. 4.1. Cylindrical coordinates.

This of course is the fully explicit representation. The fully and partially implicit forms are discussed immediately after the energy balance derivation.

4.1.3. Energy Balance

Using Figs. 4.1 and 4.2 for reference, we shall evolve the energy balance for node 0, in terms of the rate of heat flow by conduction to 0 from its

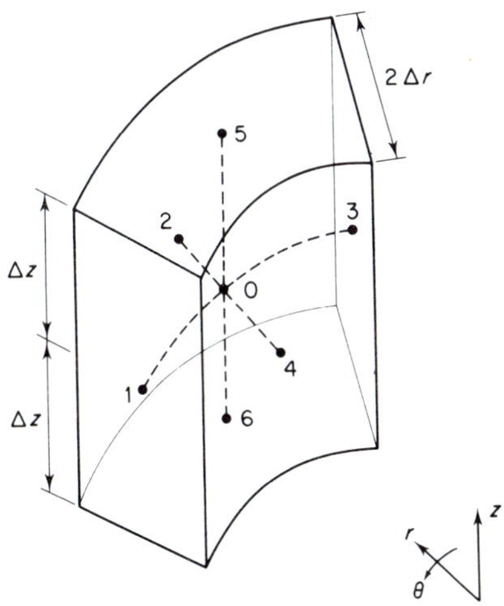

FIG. 4.2. Subdivision of cylindrical body with equal r and θ increments.

neighbouring nodes, the rate of internal heat generation in the control volume represented by 0, and the rate at which the control volume is rising in temperature:

Rate of heat flow from node 1 to node 0 by conduction $= k(\Delta z \cdot \Delta r)\left(\dfrac{T_1 - T_0}{r_0\,\Delta\theta}\right)$

Rate of heat flow from node 2 to node 0 by conduction $= k(\Delta\theta \cdot \Delta z)\left(r_0 + \dfrac{\Delta r}{2}\right)\left(\dfrac{T_2 - T_0}{\Delta r}\right)$

Rate of heat flow from node 3 to node 0 by conduction $= k(\Delta z \cdot \Delta r)\left(\dfrac{T_3 - T_0}{r_0\,\Delta\theta}\right)$

Rate of heat flow from node 4 to node 0 by conduction $= k(\Delta\theta \cdot \Delta z)\left(r_0 - \dfrac{\Delta r}{2}\right)\left(\dfrac{T_4 - T_0}{\Delta r}\right)$

Rate of heat flow from node 5 to node 0 by conduction $= k(r_0\,\Delta\theta\,\Delta r)\left(\dfrac{T_5 - T_0}{\Delta z}\right)$

Rate of heat flow from node 6 to node 0 by conduction $= k(r_0\,\Delta\theta\,\Delta r)\left(\dfrac{T_6 - T_0}{\Delta z}\right)$

(4.1.3)

Rate of internal heat generation in control volume

$$= H \times \text{Control volume}$$
$$= H \times (r_0\Delta\theta \cdot \Delta r \cdot \Delta z) \tag{4.1.4}$$

The rate at which heat is transferred to the control volume can also be represented by the equation

$$Q_v = (\text{'Mass' of control volume} \times \text{Specific heat} \times \text{Rate of change of node temperature})$$

$$= \rho(r_0\Delta\theta \cdot \Delta r \cdot \Delta z)\,C_p\left(\frac{T_0' - T_0}{\Delta t}\right) \tag{4.1.5}$$

Summing all the terms in eqns. (4.1.3) and (4.1.4), and equating this sum to the expression in eqn. (4.1.5), gives, after rearranging, the complete energy balance for node 0 as

$$\left[\frac{T_2 - 2T_0 + T_4}{(\Delta r)^2} + \frac{T_2 - T_4}{2r_0\,\Delta r} + \frac{T_1 - 2T_0 + T_3}{r_0^{\,2}\,(\Delta\theta)^2} + \frac{T_5 - 2T_0 + T_6}{(\Delta z)^2} + \frac{H}{k}\right] = \frac{(T_0' - T_0)}{\alpha \cdot \Delta t} \qquad (4.1.6)$$

This equation is exactly the same as eqn. (4.1.2), and can equally well be applied to any point, except one, within the object. The one exception is a point on an axis of symmetry, since then the term $(1/r)(\partial T/\partial r)$ becomes indeterminate ($= 0/0$) because $\partial T/\partial r = 0$ on the axis. Application of L'Hôpital's rule, however, shows that as $r \to 0$,

$$\frac{1}{r}\frac{\partial T}{\partial r} \to \frac{\partial^2 T}{\partial r^2} \qquad (4.1.7)$$

which is determinate (see Smith, 1965). As an example, for the case of 2-D steady-state heat flow with cylindrical symmetry and $H = 0$, the governing equation is

$$\frac{\partial^2 T}{\partial r^2} + \frac{1}{r}\frac{\partial T}{\partial r} = 0 \qquad (4.1.8)$$

Using eqn. (4.1.7), the above equation becomes

$$2\frac{\partial^2 T}{\partial r^2} = 0, \quad \text{at } r = 0 \qquad (4.1.9)$$

The FDE for this simply takes the temperature at $r = 0$ to be equal to that of its nearest neighbour. For steady-state non-axisymmetric ($\partial T/\partial\theta \neq 0$) problems, the temperature at $r = 0$ may be taken as the average of the values at mesh points immediately surrounding it.

If all the left-hand-side terms of eqns. (4.1.6) or (4.1.2) are evaluated at the old time-level (non-primed T's), suitable rearrangement allows T_0' to be computed directly from previous time-level temperatures at 0 and its six neighbours. This is the explicit representation. If all the left-hand-side terms are evaluated at the new time-level (primed T's), the fully implicit representation results:

$$\left[\frac{T_2' - 2T_0' + T_4'}{(\Delta r)^2} + \frac{T_2' - T_4'}{2\,r_0\,\Delta r} + \frac{T_1' - 2T_0' + T_3'}{r_0^2(\Delta\theta)^2} + \frac{T_5' - 2T_0' + T_6'}{(\Delta z)^2} + \frac{H}{k}\right] = \frac{(T_0' - T_0)}{\alpha\,\Delta t} \qquad (4.1.10)$$

In general, one has the choice of which time-level the left-hand side is evaluated at, yielding the various explicit, fully implicit and weighted average intermediate representations of the spatial derivative terms. As before, in Chapter 3, Section 3.5, each of the non-explicit representations gives an equation for T'_0 which includes other new time-level primed temperatures at nearby locations. Thus T'_0 cannot be simply computed and a set of algebraic equations must be solved in order to compute the new time-level T''s at all locations simultaneously. This problem has been discussed before and will be returned to later, during the mathematical discussion of Chapter 7, Section 7.5.

4.1.4. Logarithmic Subdivision

For 2-D steady-state heat flow with no internal heat generation, eqn. (4.1.1) is modified to

$$\frac{\partial^2 T}{\partial r^2} + \frac{1}{r}\frac{\partial T}{\partial r} + \frac{1}{r^2}\frac{\partial^2 T}{\partial \theta^2} = 0 \qquad (4.1.11)$$

It is often convenient to transform eqn. (4.1.11) using a new variable ξ, defined by

$$\xi = \log_e r \qquad (4.1.12)$$

Equation (4.1.11) simplifies to

$$\frac{\partial^2 T}{\partial \xi^2} + \frac{\partial^2 T}{\partial \theta^2} = 0 \qquad (4.1.13)$$

This is only possible if the solution domain does not include the point $r = 0$, since as $r \to 0$, $\xi \to -\infty$. Equation (4.1.13) is identical to the equation generated by the cartesian coordinate system and can be handled in the same way. Care must be taken when interpreting the answers, because equal increments in ξ do not correspond to equal increments in r. The following example clarifies this concept.

4.1.5. Example

Figure 4.3 shows the cross-section of a part cylinder. The cylinder is of 10 mm internal radius and 70 mm external radius, and the temperatures over the boundary are as shown. The problem would normally be that of determining the temperature distribution throughout the object, but in

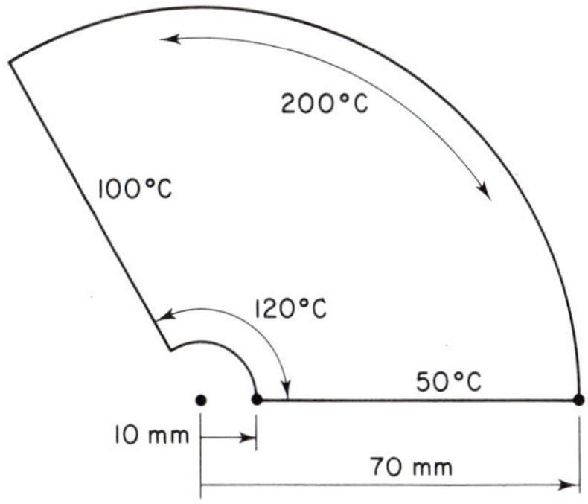

FIG. 4.3. Cylindrical section of component.

this case we shall only evolve the grid and the typical FDEs which result from the logarithmic subdivision.

$$r_i = 10\,\text{mm}$$
$$r_o = 70\,\text{mm}$$
$$\theta = 120^\circ = 2{\cdot}094\,\text{rad}$$

Applying eqn. (4.1.12):

$$\xi_o - \xi_i = \log_e \frac{r_o}{r_i} = 1{\cdot}946$$

Subdividing the θ-grid into five increments will generate an increment size of (2·094/5) rad, *i.e.*

$$\Delta\theta = 0{\cdot}419\,\text{rad}$$

Ideally, the $(\xi\text{–}\theta)$ grid would comprise all squares to facilitate the FDE derivation. If a sufficiently small grid size was chosen, it would be possible to have a square mesh. However, to be realistic, since the $\Delta\theta$ dimension is 0·419, we can investigate how many segments of size 0·419 will 'fit' into the change in ξ.

$$\text{Total change in } \xi = (\xi_o - \xi_i) = 1{\cdot}946$$

If $\Delta\xi = 0{\cdot}419$ for square grid, then the ξ-subdivision will be $(4 \times \Delta\xi + 0{\cdot}27)$, *i.e.*

$$(4 \times \Delta\xi + 0{\cdot}27) = 1{\cdot}946$$

The resulting grid appears in Fig. 4.4.

Assuming that the thermal conductivity is constant, then the FDE for any node in the square mesh region, such as node 5, will be

$$\text{Node 5: } T_2 + T_4 + T_6 + T_8 - 4T_5 = 0$$

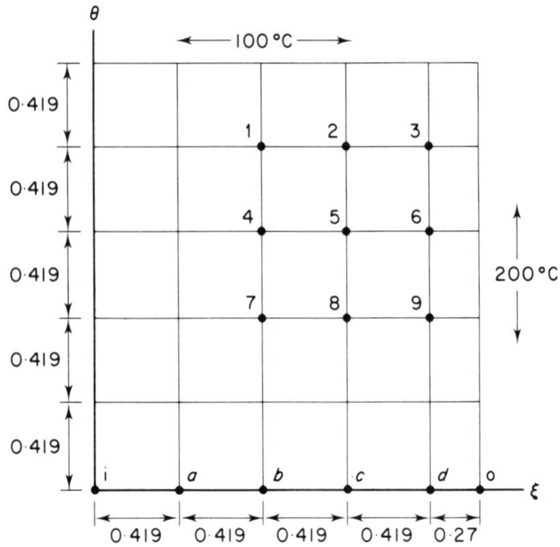

FIG. 4.4. Transformed cylindrical body using logarithmic subdivision.

For a node such as 6, which is in the irregularly shaped region (see Fig. 4.5), the FDE can be derived most easily by the energy balance method, as follows. Again assuming that the thermal conductivity is constant throughout and referring to Fig. 4.4 above, the energy balance for node 6 is

$$\left[k\left(\frac{0{\cdot}419}{2}+\frac{0{\cdot}27}{2}\right)\left(\frac{T_3-T_6}{0{\cdot}419}\right)+k\left(\frac{0{\cdot}419}{2}+\frac{0{\cdot}27}{2}\right)\left(\frac{T_9-T_6}{0{\cdot}419}\right)\right.$$
$$\left.+\,k(0{\cdot}419)\left(\frac{T_5-T_6}{0{\cdot}419}\right)+k(0{\cdot}419)\left(\frac{200-T_6}{0{\cdot}27}\right)\right]=0$$

This equation resolves to

$$\text{Node 6}: T_3 + 1{\cdot}216T_5 + T_9 - 5{\cdot}103T_6 = -377{\cdot}5 \qquad (4.1.14)$$

The locations of the corresponding points in the original system are determined by calculating the radii r_a, r_b, r_c and r_d from the values of ξ_a, ξ_b, ξ_c and ξ_d:

$$\xi_i = \log_e r_i = 2{\cdot}303 \quad \text{(which is the } \xi \text{ value at } \theta = 0 \text{ in Fig. 4.4)}$$

$$\xi_a - \xi_i = 0{\cdot}419 \;\therefore\; \xi_a = 2{\cdot}722, \text{ giving } r_a = 15{\cdot}2\text{ mm}$$
$$\xi_b - \xi_i = 0{\cdot}839 \;\therefore\; \xi_b = 3{\cdot}141, \text{ giving } r_b = 23{\cdot}1\text{ mm}$$

Similarly, $r_c = 35{\cdot}15$ mm and $r_d = 53{\cdot}44$ mm.

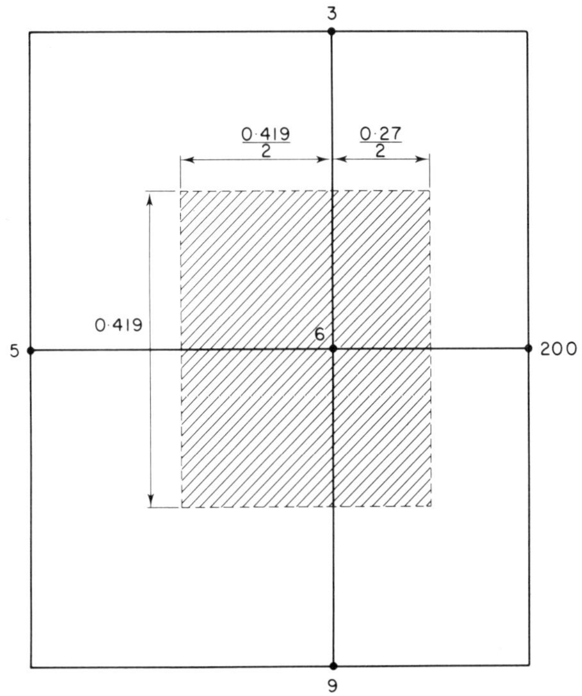

FIG. 4.5. Irregular mesh around node 6.

4.1.6. Spherical Polar Coordinates

The general equation of heat conduction in these coordinates (taking the thermal conductivity as constant) is

$$\frac{k}{r^2}\left[\frac{\partial}{\partial r}\left(r^2\frac{\partial T}{\partial r}\right)+\frac{1}{\sin\theta}\frac{\partial}{\partial\theta}\left(\sin\theta\frac{\partial T}{\partial\theta}\right)+\frac{1}{\sin^2\theta}\frac{\partial^2 T}{\partial\phi^2}\right]+H=\rho C_{\mathrm{p}}\frac{\partial T}{\partial t} \tag{4.1.15}$$

(The derivation of this equation is discussed in Section 2.3 of Chapter 2.) Figure 4.6 shows a group of six nodes centred on the point 0 with equal r, θ and ϕ increments. The FDE which replaces eqn. (4.1.15) above will be derived using the energy balance method. To deduce the appropriate equation for the temperature T_0' via this method, the six conductances between the point 0 and its six neighbours are required. Equation (3.4.3) of

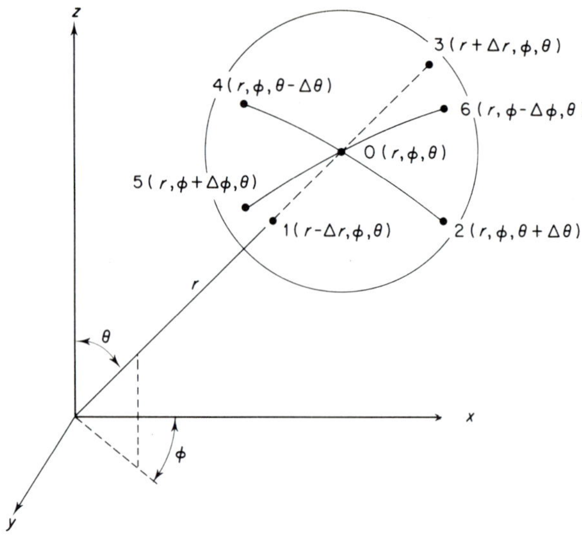

FIG. 4.6. Spherical polar coordinate system.

Chapter 3 is interpreted as

$$K_{10} = k\frac{\left(r - \frac{\Delta r}{2}\right)^2 \Delta\phi\,\Delta\theta \sin\theta}{\Delta r}$$

$$K_{20} = k\frac{r\,\Delta r\,\Delta\phi \sin\left(\theta + \frac{\Delta\theta}{2}\right)}{r\,\Delta\theta}$$

$$K_{30} = k\frac{\left(r + \frac{\Delta r}{2}\right)^2 \Delta\phi\,\Delta\theta \sin\theta}{\Delta r}$$

$$K_{40} = k\frac{r\,\Delta r\,\Delta\phi \sin\left(\theta - \frac{\Delta\theta}{2}\right)}{r\,\Delta\theta}$$

$$K_{50} = K_{60} = k\frac{r\,\Delta r\,\Delta\theta}{r\,\Delta\phi \sin\theta} \qquad (4.1.16)$$

$$\text{Control volume for node } 0 = \Delta V_0 = r^2\,\Delta\theta\,\Delta\phi\,\Delta r \sin\theta$$

Thus the energy balance for node 0 will be

$$[K_{10}(T_1 - T_0) + K_{20}(T_2 - T_0) + K_{30}(T_3 - T_0) + K_{40}(T_4 - T_0) + K_{50}(T_5 - T_0) + K_{60}(T_6 - T_0)] + H\Delta V_0 = \rho \Delta V_0 C_p \frac{(T_0' - T_0)}{\Delta t} \quad (4.1.17)$$

Equation (4.1.17) is an explicit FDE and the comments relating to the implicit derivation are exactly those previously related in this section to the cylindrical polar coordinates.

With the spherical polar coordinate system, a problem similar to that with cylindrical polars occurs if the point $r = 0$ is included in the solution domain. The term $(2/r)(\partial T/\partial r)$ may be replaced by $2(\partial^2 T/\partial r^2)$ at $r = 0$, again using L'Hôpital's rule; the terms θ and ϕ for the $r = 0$ point are omitted and the equation becomes

$$3\frac{\partial^2 T}{\partial r^2} = 0$$

Again the FDE for this simply takes the temperature at $r = 0$ equal to the mean of its immediate neighbours, but if there is a transient term involving $\partial T/\partial t$ then the full FDE must be written out.

4.1.7. Special Forms of the General FDE

In the case of problems which do not have three space coordinates and time as independent variables, reduced forms of the general governing PDE are appropriate. Tables 2.4 to 2.6 in Chapter 2 presented some of these simplified versions of the PDEs. It is now possible and convenient to give the corresponding FDEs, assuming in each case that equal mesh spacings occur in the coordinate directions. Tables 4.1 to 4.3 give these equivalent FDE simulations. One must remember, of course, that in each case a single continuous partial differential equation governing the problem is being replaced by a set of (many) algebraic finite difference equations, one such equation for the unknown temperature at each node point of the predetermined mesh system covering the region of interest.

4.2. TRIANGULAR NETWORKS

Many body shapes occur which cannot be satisfactorily fitted with the mesh systems already described: cartesian, cylindrical and spherical coordinate

TABLE 4.1

REDUCED FORM OF GENERAL FDE IN CARTESIAN COORDINATES, CORRESPONDING TO TABLE 2.4 OF CHAPTER 2

Problem	*FDE*	*Required variable*	*Comments*
2-D steady-state	$(T_O - 2T_P + T_I)/\Delta^2 + (T_E - 2T_P + T_W)/\Delta^2 = 0$	T_P	Uniform grid with $\Delta = \Delta x = \Delta y$
3-D steady-state with internal heat generation	$T_O + T_I + T_E + T_W + T_N + T_S - 6T_P + H\Delta^2/k = 0$	T_P	Uniform grid with $\Delta = \Delta x = \Delta y = \Delta z$
1-D transient	$\alpha(T_O - 2T_P + T_I)/(\Delta x)^2 = (T'_P - T_P)/\Delta t$	T'_P	Explicit representation
2-D transient with internal heat generation	$H\Delta^2/(\rho C_p) + \alpha(T_O + T_I + T_E + T_W - 4T_P) = \Delta^2(T'_P - T_P)/\Delta t$	T'_P	Uniform spatial grid with $\Delta = \Delta x = \Delta y$ Explicit representation

TABLE 4.2

REDUCED FORM OF GENERAL FDE IN CYLINDRICAL POLAR COORDINATES, CORRESPONDING TO TABLE 2.5 OF CHAPTER 2

Problem	*FDE*	*Required variable*	*Comments*
2-D steady-state Disk	$\frac{T_2 - 2T_0 + T_4}{(\Delta r)^2} + \frac{T_2 - T_4}{2r_0\,\Delta r} + \frac{T_1 - 2T_0 + T_3}{r_0^2(\Delta\theta)^2} = 0$	T_0	$\frac{\partial T}{\partial z} = 0$ Uniform grid
Cylinder	$\frac{T_2 - 2T_0 + T_4}{(\Delta r)^2} + \frac{T_2 - T_4}{2r_0\,\Delta r} + \frac{T_5 - 2T_0 + T_6}{(\Delta z)^2} = 0$		$\frac{\partial T}{\partial \theta}$
3-D steady-state with internal heat generation	$\frac{T_2 - 2T_0 + T_4}{(\Delta r)^2} + \frac{T_2 - T_4}{2r_0\,\Delta r} + \frac{T_1 - 2T_0 + T_3}{r_0^2(\Delta\theta)^2} + \frac{T_5 - 2T_0 + T_6}{(\Delta z)^2} + \frac{H}{k} = 0$	T_0	Uniform grid
1-D transient	$\alpha\left(\frac{T_2 - 2T_0 + T_4}{(\Delta r)^2} + \frac{T_2 - T_4}{2r_0\,\Delta r}\right) = \frac{T_0' - T_0}{\Delta t}$	T_0'	Uniform grid Explicit representation
2-D transient with internal heat generation Disk	$\frac{H}{\rho C_p} + \alpha\left(\frac{T_2 - 2T_0 + T_4}{(\Delta r)^2} + \frac{T_2 - T_4}{2r_0\,\Delta r} + \frac{T_1 - 2T_0 + T_3}{r_0^2(\Delta\theta)^2}\right) = \frac{T_0' - T_0}{\Delta t}$		
Cylinder	$\frac{H}{\rho C_p} + \alpha\left(\frac{T_2 - 2T_0 + T_4}{(\Delta r)^2} + \frac{T_2 - T_4}{2r_0\,\Delta r} + \frac{T_5 - 2T_0 + T_6}{(\Delta z)^2}\right) = \frac{T_0' - T_0}{\Delta t}$	T_0'	Uniform grid Explicit representation

TABLE 4.3

REDUCED FORM OF GENERAL FDE IN SPHERICAL POLAR COORDINATES, CORRESPONDING TO TABLE 2·6 OF CHAPTER 2

Problem	*FDE*	*Required variable*	*Comments*
2-D steady-state	Eqn. (4.1.17) with 5th, 6th and 7th terms on LHS omitted and RHS term omitted Eqn. (4.1.17) with 2nd, 4th and 7th terms on LHS omitted and RHS term omitted	T_0	Uniform grid
3-D steady-state with internal heat generation	Eqn. (4.1.17) with RHS term omitted	T_0	Uniform grid
1-D transient	Eqn. (4.1.17) with 2nd, 4th, 5th, 6th and 7th terms on LHS omitted	T_0'	Uniform grid Explicit representation
2-D transient with internal heat generation	Eqn. (4.1.17) with 5th and 6th terms on LHS omitted Eqn. (4.1.17) with 2nd and 4th terms on LHS omitted	T_0'	Uniform grid Explicit representation

types. Moreover, in many practical problems, though a usual square or rectangular net is suitable for the interior and some of the boundary of a body, any irregular parts of the boundary (angled or curved, for example) require special treatment. It is then often convenient to use a triangular network, either throughout or locally. The energy balance technique is most suitable for deriving an appropriate FDE for each point of the mesh, and the thermal conductances between nodes can be simply calculated by a unified method (see Dusinberre, 1961; Razelos, 1973).

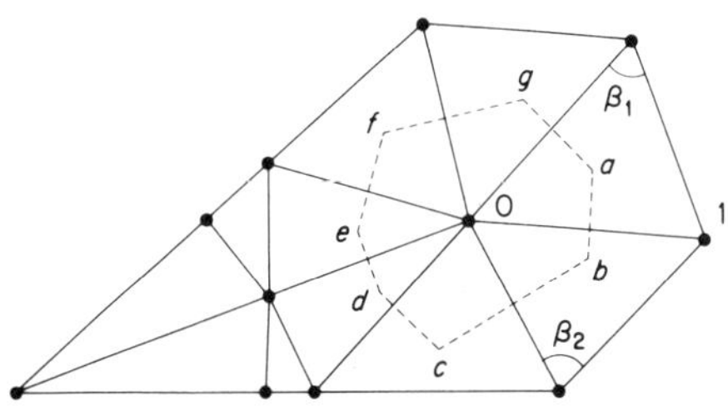

FIG. 4.7. Part of a general irregular triangular mesh system. Control area (shown dotted) = *abcdefg*.

Part of a general irregular triangular mesh system is shown in Fig. 4.7, where all internal angles are acute ($\leq 90°$). This acute restriction is necessary in order that the control volume (area for the case illustrated) for the point labelled 0 may have perpendicular bisectors (of the lines joining the point 0 to each of its neighbours) as its sides, as illustrated. It is then possible to calculate the thermal conductances between the node 0 and each of its neighbours as follows.

Interpretation of eqn. (3.4.3) of Chapter 3 for the thermal conductance between nodes 0 and 1, for example, is

$$K_{01} = \frac{kL_{\text{ab}}}{L_{01}}$$

Using simple geometry this equation becomes

$$K_{01} = \frac{k(\cot\beta_1 + \cot\beta_2)}{2} \tag{4.2.1}$$

which is sometimes written

$$K_{01} = kF_{01}$$

where F_{01} is called the geometric factor between nodes 0 and 1. There are six other neighbours of the node 0 shown in the figure whose conductances are evaluated similarly. One can then set up the FDE for the node 0 using eqn. (3.4.2) of Chapter 3, repeated here for convenience in the current notation and assuming steady-state conditions:

$$\sum_j K_{0j}(T_j - T_0) + H \cdot V_0 = 0 \tag{4.2.2}$$

The volume V_0 associated with the node 0 is the area of the surrounding polygon, shown dotted in the figure, and is given by

$$V_0 = \tfrac{1}{2} L_{ab} \frac{L_{01}}{2} + \text{six other similar terms} \tag{4.2.3}$$

This volume will also be required if transient conditions exist, for then the term

$$\rho C_p V_0 \frac{\Delta T}{\Delta t}$$

will replace the right-hand side of eqn. (4.2.2).

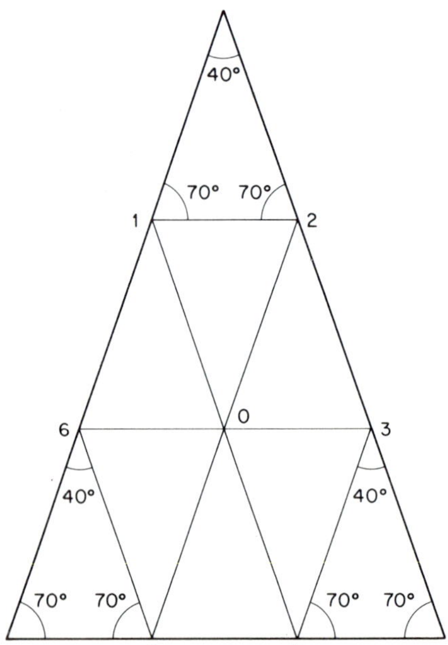

FIG. 4.8. Triangular net for an isosceles triangle.

4.2.1. Example

Bodies of triangular shape are conveniently subdivided into similar triangular sections. Figure 4.8 shows an isosceles triangle-shaped section with all angles either 70° or 40°.

The thermal conductances between node 0 and the surrounding nodes 1, 2, 3, 4, 5 and 6 are derived by using eqn. (4.2.1):

$$K_{01} = 2 \times (\cot 70°)/2 \quad \text{(contribution from triangle 120}$$
$$= 0{\cdot}364 \quad \text{plus contribution from triangle 160)}$$

$$K_{02} = K_{04} = K_{05} = K_{01} \quad \text{(by symmetry)}$$

$$K_{03} = K_{06} = 2 \times (\cot 40°)/2 = 1{\cdot}192$$

Thus for a steady-state 2-D problem with no internal heat generation, the energy balance yields the following equations for the node 0:

$$0{\cdot}364(T_1 - T_0) + 0{\cdot}364(T_2 - T_0) + 0{\cdot}364(T_4 - T_0) + 0{\cdot}364(T_5 - T_0)$$
$$+ 1{\cdot}192(T_3 - T_0) + 1{\cdot}192(T_6 - T_0) = 0$$

i.e.
$$T_1 + T_2 + T_4 + T_5 + 3{\cdot}274(T_3 + T_6) - 10{\cdot}549T_0 = 0 \qquad (4.2.4)$$

4.2.2. Boundary Nodes

Boundary nodes, whose temperatures are unknown and implied in a convection condition, can be treated in a similar way using a local triangular mesh, even if other parts of the body do not have such a mesh. For example, a portion of a boundary is shown in Fig. 4.9 and perpendicular bisectors are shown dotted. To set up an equation for the node 0, the control volume

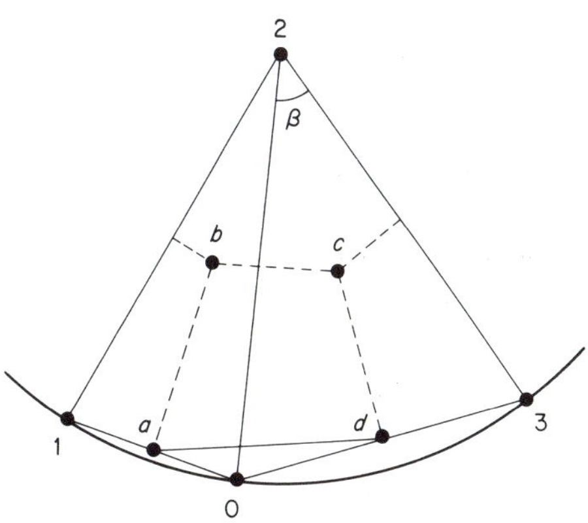

FIG. 4.9. Boundary node of irregular triangular mesh.

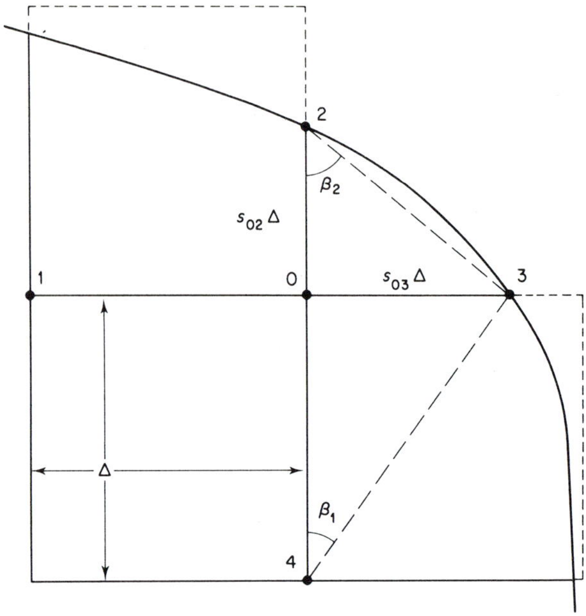

FIG. 4.10. Square mesh terminating in irregular boundary.

(area) is *abcd*. A typical thermal conductance between nodes 0 and 3 is

$$K_{03} = k\frac{\cot \beta}{2}$$

The node 0 is acted upon by the nodes 1, 2 and 3 and the ambient. To satisfy this last one (a boundary condition which involves specified heat flux, Newtonian cooling on derivative boundary conditions), the required surface element is taken equal to L_{ad}.

Square meshes terminating in irregular boundaries such as the one shown in Fig. 4.10 are easily handled with this method. Thus, in order to set up the FDE for the node 0 and assuming known boundary conditions, the conductances must first be derived:

$$K_{03} = k\left(\frac{\cot \beta_1 + \cot \beta_2}{2}\right)$$

$$= \frac{k}{2}\left(\frac{1}{s_{03}} + \frac{s_{02}}{s_{03}}\right) = \frac{k}{2s_{03}}[1 + s_{02}]$$

Similarly

$$K_{02} = \frac{k}{2s_{02}}(1 + s_{03})$$

$$K_{01} = \frac{k}{2}(1 + s_{02})$$

$$K_{04} = \frac{k}{2}(1 + s_{03})$$

Rectangular meshes terminating in irregular boundaries can be handled similarly. The result will accord with eqn. (3.3.3) of Chapter 3.

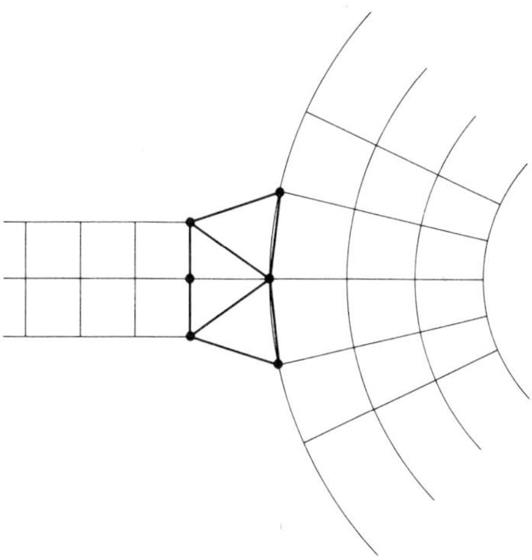

FIG. 4.11. Interconnection of two different meshes.

Often it is advantageous to use a different mesh shape for different parts of a problem domain. For example, if more accuracy is required in some regions or if strong variations in temperature are expected in some regions, then these regions should be given a finer mesh. Consider a 2-D cartesian steady-state problem covered with a uniform square or rectangular grid. If a sudden change to a different mesh shape is required, the principles of triangular meshes can be used. Figure 4.11 illustrates how a square grid can be merged with a grid representing cylindrical polar coordinates. An illustration of the full procedure involved with triangular meshes is given in Chapter 8, Section 8.5.

4.3. BOUNDARIES, INTERFACES, COMPOSITE MEDIA AND NON-UNIFORM PROPERTIES

4.3.1 Boundary Nodes

So far only slight mention has been made of boundary nodes and the incorporation of boundary conditions appropriate to the problem, in Section 3.4 of Chapter 3 and Sections 4.1 and 4.2 of the current chapter. Consider a problem which has a grid system covering the solution domain with boundaries coinciding with mesh lines. In order to be soluble a FDE must be provided for each one of the entire set of node points, which in general consists of all internal points and all boundary points. The FDE for each internal point can be deduced from the material presented so far; the objective here is to supply an equation for each boundary point.

In heat transfer problems two cases often arise for each boundary point, of which the point B in Fig. 4.12 is an example:

(i) Boundary surface temperature is known (Dirichlet known boundary condition), in which case

$$T_B = T_G \tag{4.3.1}$$

where T_G is the known temperature given in the problem specification. Thus eqn. (4.3.1) provides a simple FDE for the point B; without it, T_B would occur in the set of unknowns and the problem would not be closed.

(ii) Boundary surface temperature is not known (Neumann derivative boundary condition), in which case the additional problem specification is in the form of knowledge about the outward normal gradient of temperature $\partial T/\partial n$. This may be:

$$\frac{\partial T}{\partial n} = K \quad \text{(given)} \tag{4.3.2}$$

where K is simply a given value;

$$\frac{\partial T}{\partial n} = -\frac{h}{k}(T_B - T_A) \quad \text{(convective)} \tag{4.3.3}$$

where h is the surface heat transfer coefficient (which is usually taken to be known beforehand but is actually dependent on the circumstances in question—see Chapter 2, Section 2.1), k is conductivity, and T_B and T_A are boundary and ambient temperatures, respectively. Equation (4.3.2) may also take the form

$$\frac{\partial T}{\partial n} = -\frac{FE\sigma}{k}(T_B^4 - T_A^4) \quad \text{(radiative)} \tag{4.3.4}$$

where F is a geometric factor, E is emissivity and σ is the Stefan–Boltzmann constant. (The multiple $F \cdot E$ is often replaced by a single letter $\mathscr{F}$ as a combined configuration-emissivity factor.)

Given these two general cases, the problem of how to incorporate eqns. (4.3.1) to (4.3.4) into the FDEs for the boundary nodes will now be discussed. Equation (4.3.1) in fact presents no new problem, since if the boundary temperatures are known, they can be inserted directly into the relevant FDEs. The other boundary conditions—convective and radiative—will be considered separately, in each case, for steady-state and transient conditions.

4.3.2. Convective Boundary Conditions (Steady-State)

The convective boundary condition, given by eqn. (4.3.3), will now be considered, with reference to Fig. 4.12, which shows part of a rectangular 2-D cartesian coordinate body adjoining an environment at a temperature T_A. For simplicity, the grid points will be taken to lie on a uniform square mesh of side Δ, and internal heat generation will be ignored. The FDEs can of course be derived by two methods, the direct PDE replacement and the energy balance method.

(i) *PDE replacement*. This method requires the introduction of a 'fictitious' node point F (see Fig. 4.12), at a distance Δ from node B. One cannot assume $T_F = T_A$, however, since T_F is affected by the proximity of

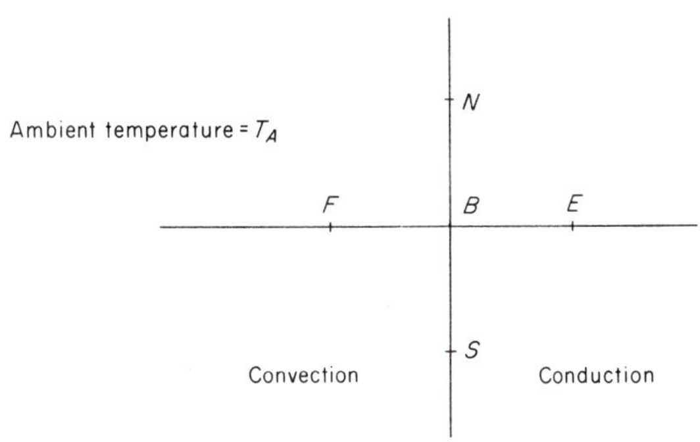

FIG. 4.12. Western convective boundary surface of a 2-D cartesian body with a square mesh.

the body, whereas T_A is the ambient temperature far from the body. The usual FDE for the point B is applied, which introduces the temperature at the fictitious point F. This is then eliminated via a central difference approximation to $\partial T/\partial n$ at B and eqn. (4.3.3). Thus

$$T_N + T_S + T_E + T_F - 4T_B = 0$$

$$\frac{T_F - T_E}{2\Delta} = -\frac{h}{k}(T_B - T_A) \qquad (4.3.5)$$

These two equations suffice to determine T_B. Usually T_F is eliminated between them, so obtaining an equation for T_B, as required.

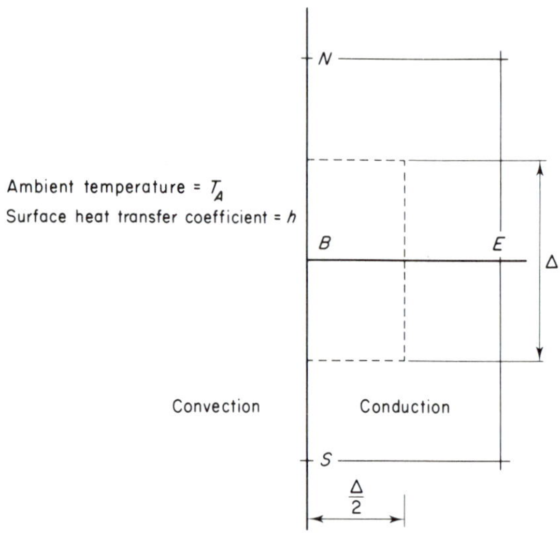

FIG. 4.13. Control volume for point B on a western convective boundary surface of a 2-D cartesian body with a square mesh.

(ii) *Energy balance method.* Figure 4.13 shows how a cell control volume (area) is set up for the boundary point B, and how T_B is affected by the ambient temperature T_A by way of a heat transfer coefficient h. As discussed in Section 3.4 of Chapter 3 and repeated here for convenience, the conductances K_{ij} between the node i and its adjacent nodes j are calculated and then the rate of heat flow Q_{ij} (W = J/s) from j to i is

$$Q_{ij} = K_{ij}(T_j - T_i) \qquad (4.3.6)$$

for each of the adjacent nodes j. The heat balance for the node i gives the following equation:

$$\sum_j Q_{ij} = \sum_j K_{ij}(T_j - T_i) = 0 \tag{4.3.7}$$

assuming no internal heat generation and steady state. For internal nodes j,

$$K_{ij} = \frac{kA_{ij}}{L_{ij}} \tag{4.3.8}$$

where k is thermal conductivity, A_{ij} is the average surface area perpendicular to the direction between j and i, and L_{ij} is the distance between the nodes i and j. The conductance from an ambient node j is

$$K_{ij} = h_i A_{ij} \tag{4.3.9}$$

where h_j is the average surface heat transfer coefficient on the surface of area A_{ij} (of the cell control volume for point i) exposed to the environment at temperature T_j. For the case considered here, i is taken as B and the four j's are N, S, E and A (ambient). The L_{ij}'s and A_{ij}'s are simply computed from the cell control volume for B, shown dotted in Fig. 4.13. Interpreting eqn. (4.3.7) for this case, and using eqns. (4.3.8) and (4.3.9) en route, yields

$$k \cdot \frac{\Delta}{2} \cdot \frac{T_N - T_B}{\Delta} + k \cdot \frac{\Delta}{2} \cdot \frac{T_S - T_B}{\Delta} + k \cdot \Delta \cdot \frac{T_E - T_B}{\Delta} + h\Delta(T_A - T_B) = 0 \tag{4.3.10}$$

from which rearrangement gives

$$T_B = \frac{\frac{1}{2}(T_N + T_S) + T_E + \mathrm{Bi}\, T_A}{2 + \mathrm{Bi}} \tag{4.3.11}$$

where $\mathrm{Bi} = (h/k)\Delta$ is the Biot number. Elimination of T_F in eqn. (4.3.5) yields this same equation for T_B, showing the equivalence of the two techniques in this case.

Equation (4.3.11) is not appropriate, however, if the boundary point B occurs not on a side but on a corner, as shown in Fig. 4.14 for the case of a south-west corner. Interpreting the heat balance approach, eqn. (4.3.7), with the help of eqns. (4.3.8) and (4.3.9), for the boundary point B gives

$$k\frac{\Delta}{2} \cdot \frac{T_N - T_B}{\Delta} + k\frac{\Delta}{2} \cdot \frac{T_E - T_B}{\Delta} + h\left(\frac{\Delta}{2} + \frac{\Delta}{2}\right)(T_A - T_B) = 0 \tag{4.3.12}$$

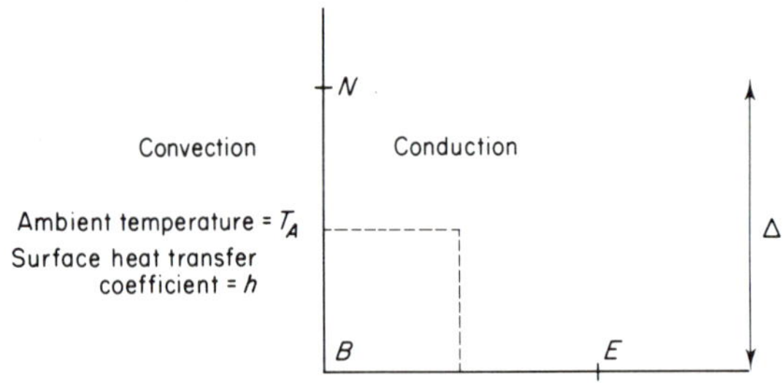

FIG. 4.14. Control volume for point B on a south-west corner convective boundary of a 2-D cartesian body with a square mesh.

using the fact that a surface area of $(\frac{1}{2}\Delta + \frac{1}{2}\Delta)$ is exposed to the ambient at temperature T_A. Rearrangement, and substitution for the Biot number Bi $= (h/k)\Delta$, yields

$$T_B = \frac{\frac{1}{2}(T_N + T_E) + \text{Bi}\, T_A}{1 + \text{Bi}} \tag{4.3.13}$$

This expression could also have been deduced by direct finite difference approximation to the governing PDE at B, with the introduction and elimination of two fictitious points.

4.3.3. Convective Boundary Condition (Transient)

Generally, if there is a heat generation term or transient conditions, equivalence of the two methods is also observed. In this connection, it may be noted that simple modifications to eqns. (4.3.5) and (4.3.10) can extend the ideas presented. Consider, for example, the 1-D transient problem with constant properties ρ, C_p and k, subject to convective boundary condition at its left-hand edge, as shown in Fig. 4.15. Over a small time-interval Δt, the

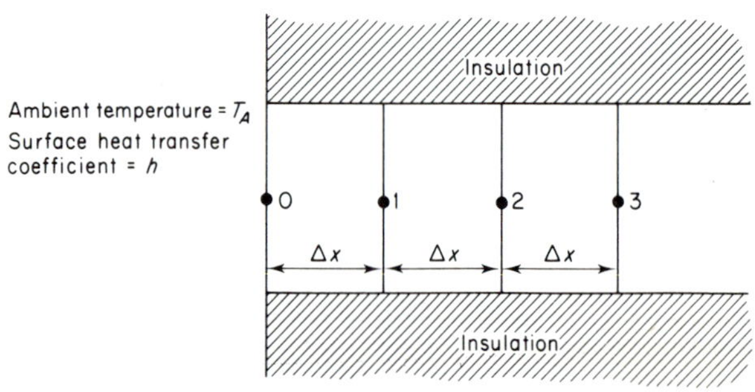

FIG. 4.15. 1-D convective boundary surface.

energy balance method gives the following equation for the boundary node labelled 0:

$$\rho C_p V_0 \frac{\Delta T_0}{\Delta t} = \frac{k}{\Delta x}(T_1 - T_0) + h(T_A - T_0) \qquad (4.3.14)$$

where $V_0 = \Delta x/2$.

Equation (4.3.14) is written explicitly, in that the temperatures on the right-hand side of the equation are already known (at time t). The equation could equally well be expressed implicitly, with the temperatures on the right expressed as those at time $t + \Delta t$. In accordance with previous discussion in Section 3.5 of Chapter 3, the explicit case is said to use backward-time differences, and the implicit case forward-time differences. It is of course also possible to use a weighted average of forward- and backward-time differences. Thus eqn. (4.3.14) can be rearranged and expressed in the explicit form

$$T_0' = 2\,\mathrm{Fo}\left[T_1 + \mathrm{Bi}\cdot T_A + T_0\left(\frac{1}{2\,\mathrm{Fo}} - 1 - \mathrm{Bi}\right)\right] \qquad (4.3.15)$$

The implicit form is

$$T_0 = 2\,\mathrm{Fo}\left[-T_1' - \mathrm{Bi}\,T_A' + T_0'\left(\frac{1}{2\,\mathrm{Fo}} + 1 + \mathrm{Bi}\right)\right] \qquad (4.3.16)$$

where $\mathrm{Fo} = \dfrac{k\,\Delta t}{\rho C_p(\Delta x)^2}$ and $\mathrm{Bi} = \dfrac{h\,\Delta x}{k}$

In a 2-D rectangular region with a square mesh of side Δ, two types of boundary nodes occur as portrayed in Fig. 4.16, one on a side (*e.g.* node 1)

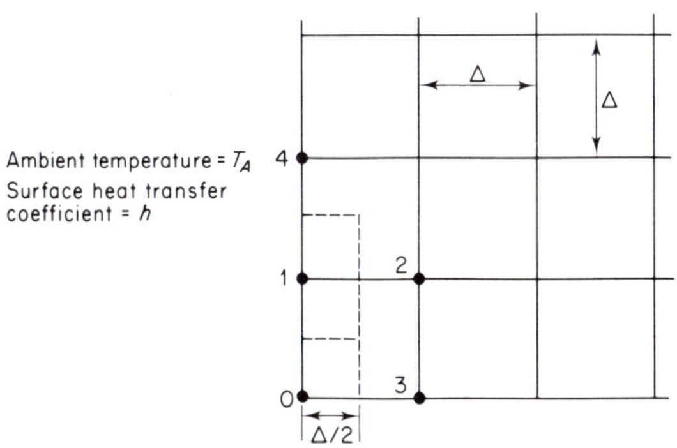

FIG. 4.16. 2-D convective boundary surface.

and one on a corner (*e.g.* node 0). The implicit and explicit FDEs for nodes 0 and 1 are stated below.

For node 0:

(Explicit)

$$T_0' = 2\,\mathrm{Fo}\left[T_1 + T_3 + 2\,\mathrm{Bi}\,T_A + T_0\left(\frac{1}{2\,\mathrm{Fo}} - 2 - 2\,\mathrm{Bi}\right)\right] \qquad (4.3.17)$$

(Implicit)

$$T_0 = 2\,\mathrm{Fo}\left[-T_1' - T_3' - 2\,\mathrm{Bi}\,T_A' + T_0'\left(\frac{1}{2\,\mathrm{Fo}} + 2 + 2\,\mathrm{Bi}\right)\right] \qquad (4.3.18)$$

For node 1:

(Explicit)

$$T_1' = \mathrm{Fo}\left[T_0 + 2T_2 + T_4 + 2\,\mathrm{Bi}\,T_A' + T\left(\frac{1}{\mathrm{Fo}} - 4 - 2\,\mathrm{Bi}\right)\right] \qquad (4.3.19)$$

(Implicit)

$$T_1 = \mathrm{Fo}\left[-T_0' - 2T_2' - T_4 - 2\,\mathrm{Bi}\,T_A' + T_1'\left(\frac{1}{\mathrm{Fo}} + 4 + \mathrm{Bi}\right)\right] \qquad (4.3.20)$$

Notice that the implicit formulations contain several unknowns at the new time-level, adding to the complexity of solution, whereas the explicit formulations generate new values of temperature immediately.

4.3.4. Radiative Boundary Condition (Steady-State)

The incorporation of eqn. (4.3.4) representing radiative boundary conditions is not quite as simple as eqn. (4.3.3) was for convective boundary conditions. Suppose, for example, that the 2-D steady-state problem just described and shown in Fig. 4.13 has a radiative boundary condition applicable at B, instead of the convective one. The energy balance for B is

$$k\frac{\Delta}{2}\cdot\frac{T_N - T_B}{\Delta} + k\frac{\Delta}{2}\cdot\frac{T_S - T_B}{\Delta} + k\,\Delta\cdot\frac{T_E - T_B}{\Delta} + \mathscr{F}\sigma(T_A^4 - T_B^4) = 0 \qquad (4.3.21)$$

which unfortunately is non-linear in its principal unknown T_B. Anticipating our relaxation solution procedure, it is convenient to linearise the above equation by using the approximation

$$T_B^4 \approx T_{B,0}^3 \cdot T_B \qquad (4.3.22)$$

where $T_{B,0}$ is the 'old' and therefore known value at node B, thus treating part of T_B^4 as a coefficient and part as an unknown T_B. Alternatively, taking $\phi = T^4$ will generate the approximation

$$\phi \approx \phi_0 + \left.\frac{\partial \phi}{\partial T}\right|_0 \Delta T$$

so that

$$\begin{aligned} T_B^4 &\approx T_{B,0}^4 + 4T_{B,0}^3(T_B - T_{B,0}) \\ &= 4T_{B,0}^3 T_B - 3T_{B,0}^4 \end{aligned} \tag{4.3.23}$$

which again is linear in T_B, and uses the 'old' value at node B as a coefficient together with a constant term.

4.3.5. Radiative Boundary Condition (Transient)

These ideas may also be extended to transient conditions. Referring again to the 1-D object shown in Fig. 4.15 but now subject to radiative heat flow at the boundary, the energy balance over a small time Δt yields the following:

$$\rho C_p \frac{\Delta x}{2}\frac{\Delta T_0}{\Delta t} = \frac{k}{\Delta x}(T_1 - T_0) + \mathscr{F}\sigma(T_A^4 - T_0^4) \tag{4.3.24}$$

The explicit and implicit formulations are therefore

(Explicit) $$T_0' = 2\,\mathrm{Fo}\left[T_1 + \left(\frac{1}{2\,\mathrm{Fo}} - RT_0^3 - 1\right)T_0 + RT_A^4\right] \tag{4.3.25}$$

(Implicit)

$$\frac{1}{2\,\mathrm{Fo}}T_0 + 3R(T_0^3 - T_A^3) = \left(\frac{1}{2\,\mathrm{Fo}} + 1 + 4RT_0^3\right)T_0' - T_1' - 4RT_A^3 T_A' \tag{4.3.26}$$

where $R = \mathscr{F}\sigma\Delta x/k$, which is analogous to the Biot number. Equation (4.3.23) has been used to suppress the non-linearity in the implicit equation. Section 8.9 in Chapter 8 illustrates the points covered.

4.3.6. Irregular Boundaries

Many physical problems have irregular boundaries, for example curved boundaries with cartesian coordinates as shown in Fig. 4.17. Consider the problem to be 2-D steady-state with no internal heat generation and covered with a regular rectangular mesh with spacings Δx and Δy. Given any $\Delta > 0$, take s_W and s_S such that

$$s_W\Delta = \Delta x \quad \text{and} \quad s_S\Delta = \Delta y$$

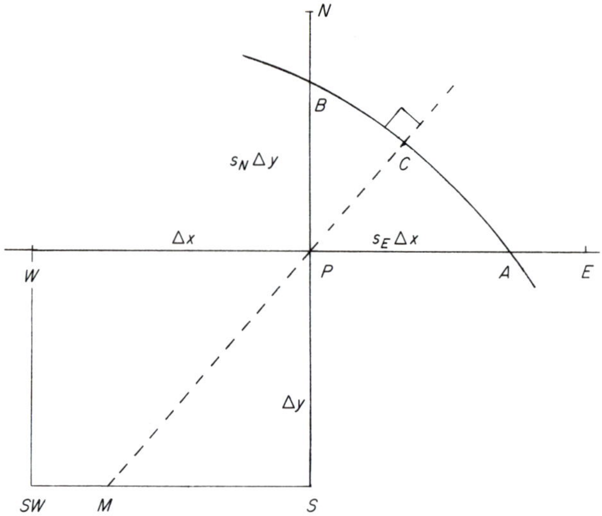

FIG. 4.17. Finite differences near a curved boundary.

If the boundary condition is one of known temperature (Dirichlet condition), then T_A and T_B are known, where A and B are points at the intersection of the curved boundary and the usual grid lines. The FDE for the point P will be derived from the irregular mesh details already presented in Section 3.3 of Chapter 3. The appropriate form of Laplace's equation, centred at P with s_N and s_E as in the figure, is

$$\frac{2}{(\Delta x)^2}\left[\frac{1}{s_E(s_E+1)}T_A - \frac{1}{s_E}T_P + \frac{1}{1(s_E+1)}T_W\right] + \frac{2}{(\Delta y)^2}\left[\frac{1}{s_N(s_N+1)}T_B - \frac{1}{s_N}T_P + \frac{1}{1(s_N+1)}T_S\right] = 0 \quad (4.3.27)$$

which now requires rearrangement to isolate T_P. The formula is only of order $\Delta x + \Delta y$, rather than $(\Delta x)^2 + (\Delta y)^2$ which is appropriate to the usual form in a regular rectangular grid. Hence the simulation of a curved boundary of known temperature loses an order of accuracy, and the error will propagate inwards somewhat. One could make use of a higher-order accuracy formula, which could be deduced by the method of undetermined coefficients, for example, and involve more interior points of the mesh. However, the s_N and s_E are different for different near-boundary points and a different formula would be required at all such points. Thus one usually settles for the less accurate version as given above.

For curved boundaries where Neumann conditions of normal derivative of temperature $\partial T/\partial n$ are given, the choice is either for a simple but inaccurate formula of order $\Delta x + \Delta y$ or for a more complex but more accurate formula of order $(\Delta x)^2 + (\Delta y)^2$. One simple way (Bayley *et al.*, 1972; Razelos, 1973) is to construct the normal to the boundary which passes through P and locate the point M where it cuts one of the lines S–SW or W–SW. Points S, SW and W are given usual FDEs involving P. However, T_M is deduced by linear interpolation from S to SW; and hence T_P is deduced from

$$\frac{T_P - T_M}{L_{PM}} = \left.\frac{\partial T}{\partial n}\right|_C$$

where L_{PM} is the length from P to M, and the given normal gradient is the value given or deducible from an equation like eqn. (4.3.3). Another simple way is to use a heat balance technique with a local triangular mesh system, as discussed in Section 3.2 and illustrated in Fig. 4.9. For more accurate representation of order $(\Delta x)^2 + (\Delta y)^2$ to derivative boundary conditions with curved boundaries, recourse must be made to the literature (see Ames, 1969).

4.3.7. Interfaces

As discussed in Section 1.2 of Chapter 1, heat transfer is often occurring through a composite structure made of a number of different materials, joined by a thin interface. Figure 4.18 illustrates the case of steady conduction across such an interface, where there is an apparent temperature discontinuity. The fictitious interface temperature drop ΔT_{int} is found by extrapolation of the local linear temperature profiles in each

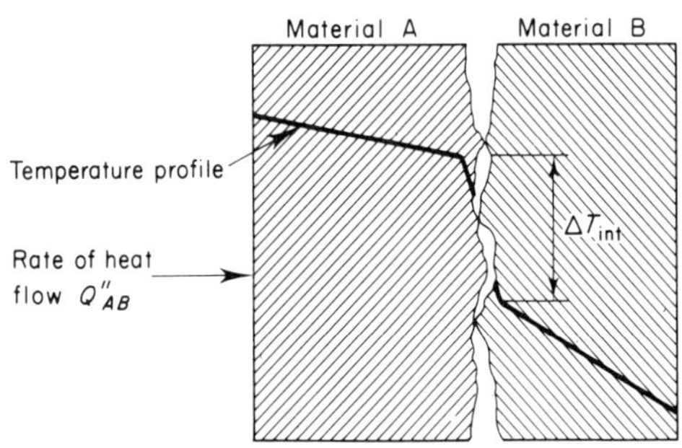

FIG. 4.18. Temperature profile across a typical interface.

material just away from the contact disturbance. With steady heat flux

$$Q''_{AB} = -k_A \frac{\partial T}{\partial x}\bigg|_A = -k_B \frac{\partial T}{\partial x}\bigg|_B \tag{4.3.28}$$

and the unit interface conductance is

$$h_i = \frac{Q''_{AB}}{\Delta T_{\text{int}}} \tag{4.3.29}$$

When there is perfect contact (sweated or welded materials, for example), the temperature drop ΔT_{int} vanishes and $h_i \to \infty$; then there is temperature continuity $T_A = T_B$ at the interface but usually a discontinuity in temperature gradient, since $k_A \neq k_B$ in general.

FDEs are derived in exactly the same way as the energy balance method was applied to the boundary nodes with convective conditions (generating eqns. (4.3.13) and (4.3.16), for example).

4.3.8. Composite Media

FDEs for situations involving heat flow through different materials can be derived by the energy balance method as before. By way of revision, we shall first look at the heat flow through a structure comprising two materials of the same dimensions but different thermal properties (see Fig. 4.19). The two materials are subjected to the same temperatures T_1 and T_2 and one-dimensional heat flow is assumed. It is a simple matter to show that the overall thermal conductivity k_{ov} is given by

$$k_{\text{ov}} = \tfrac{1}{2}(k_A + k_B) \tag{4.3.30}$$

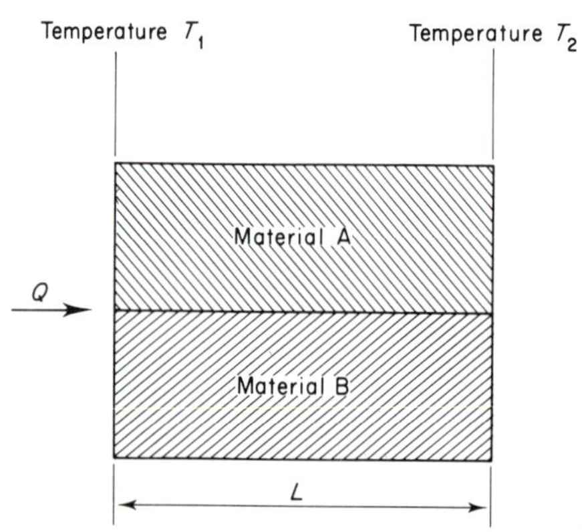

FIG. 4.19. 1-D composite system.

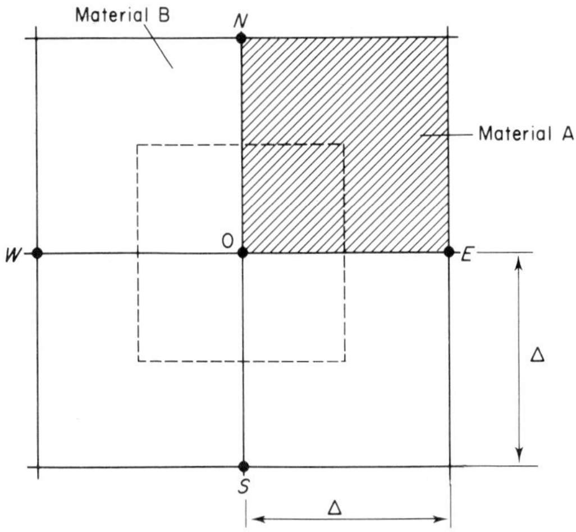

FIG. 4.20. 2-D composite system for node 0.

We shall now apply eqn. (4.3.30) directly to the 2-D problem illustrated in Fig. 4.20. The energy flow between node E and node 0 is given by the equation

$$Q_{E0} = \tfrac{1}{2}(k_A + k_B)\Delta \cdot \frac{(T_E - T_0)}{\Delta} \tag{4.3.31}$$

A similar equation will apply between nodes N and 0.

The control volume now comprises materials of different density and specific heat and in transient problems an overall (ρC_p) term must be evolved. Remembering that one-quarter of the control volume is of material A and the rest of material B, the equivalent (ρC_p) term is given by the equation

$$(\rho C_p)_{\text{node }0} = 0{\cdot}75(\rho_B C_{pB}) + 0{\cdot}25(\rho_A C_{pA}) \tag{4.3.32}$$

Applying eqns. (4.3.31) and (4.3.32) to the 2-D problem in Fig. 4.20 (ignoring internal heat generation), the FDE for node 0 under transient conditions will be

$$\left[\frac{k_A + k_B}{2}(T_N - T_0) + \frac{k_A + k_B}{2}(T_E - T_0) + k_B(T_S - T_0) + k_B(T_W - T_0)\right]$$
$$= [0{\cdot}75(\rho_B C_{pB}) + 0{\cdot}25(\rho_A C_{pA})]\,\Delta^2\,\frac{(T_0' - T_0)}{\Delta t} \tag{4.3.33}$$

Further details of this topic are included in Sections 8.9 and 8.10 of Chapter 8.

4.3.9. Non-Uniform Properties

For slight temperature variations, thermal conductivities are considered to be constant, but if large temperature variations are expected the temperature dependence of thermal conductivity k must be taken into account. For example, one has

$$k = k(T)$$

and the thermal conductance K_{ij} between two nodes i and j, distant L_{ij} apart with surface area A_{ij} normal to flow direction, is

$$K_{ij} = \frac{k(T_i) + k(T_j)}{2} \cdot \frac{A_{ij}}{L_{ij}} \tag{4.3.34}$$

If k varies linearly with T, then one can use

$$k\left(\frac{T_i + T_j}{2}\right) = \frac{k(T_i) + k(T_j)}{2}$$

In solving a problem, an average value of k may be used in order to compute the temperature field. The new k's can be found based on this new temperature field, and hence one can re-compute the temperature field. Iteration on the k's and T's is continued until satisfactory convergence is obtained. There is further discussion of this topic in Sections 8.9 and 8.10 of Chapter 8.

4.4. ACCURACY AND EXTRAPOLATION*

4.4.1. Unique Solution (see Young, 1962)

Many problems are concerned with 2-D steady-state situations and to fix our ideas, consider such a case in cartesian coordinates. The general FDE may be written in the form

$$a_P T_P = \sum_{j = N, S, E, W} a_j T_j + s_P \tag{4.4.1}$$

where the a_P and a_j's are all positive and where

$$a_P \geq a_N + a_S + a_E + a_W \tag{4.4.2}$$

Strict inequality holds when P is adjacent to a Dirichlet boundary, and then at least one of the terms of the summation is transferred to the source term s_P. For example, in a square mesh of size Δ, eqn. (3.3.2) of Chapter 3 is

$$T_O + T_I + T_E + T_W - 4T_P + \frac{H}{k}\Delta^2 = 0 \tag{4.4.3}$$

* This section is concerned with steady-state problems.

which conforms with a slight change of notation for the nodes to the form of eqn. (4.4.1) with

$$a_P = 4 \qquad a_j = 1 \text{ (for all } j) \qquad s_P = \frac{H}{k}\Delta^2$$

Most problems encountered also have a solution domain that is connected, in the sense that any two points can be connected by lines joining adjacent points in the domain.

It is possible to show that eqn. (4.4.1) holding at all points of such a domain has a unique solution. To do this it is necessary and sufficient to show that the determinant of the related linear system does not vanish; and this is so if the homogeneous system, obtained by letting boundary conditions and s_P vanish, has only the trivial solution, which vanishes everywhere. The proof is by contradiction.

Suppose this homogeneous system has a non-trivial solution; then for some point P, $T_P \neq 0$ and without loss of generality one can take $T_P > 0$. Let M be the largest such positive value of T and suppose it occurs at the point P. From eqns. (4.4.1) and (4.4.2) it follows that T must assume this value M at each of its adjacent points, and continuing this process and using the connectedness of the region it follows that $T = M$ at all points of the region, including the boundary points. This contradicts the assumption that T vanishes on the boundary, and hence proves that T vanishes everywhere in the region. It then follows that eqn. (4.4.1) has a unique solution.

4.4.2. Approximate Solution and Extrapolation

The boundary value problem may have a unique solution, but a major problem is seeking it and our knowledge of how near our approximate solution of FDEs like eqn. (4.4.1) is to the true solution of the governing PDE. Under rather general conditions it can be shown that as Δ tends to zero, the solution of the FDEs tends to the solution of the PDE. One also has the problem of how near one can get, computationally, to the exact solution of the FDEs; in addition, one has the problem of how accurate the FDE exact solution is for a given grid size Δ.

As to the former, that a relaxation method gives an approximate solution of the algebraic FDEs, an estimate of the error for a square mesh of size Δ using the five- and nine-point finite difference formulae (see Chapter 3, Section 3.3) is available. The errors do not exceed the values $RL^2\Delta^2/4$ and $RL^2\Delta^2/24$ respectively, where R is the maximum absolute residual and L is the radius of a circle enclosing the entire region (Walsh, 1966).

Bounds for the maximum error may be found in terms of local error (see

Young, 1962). For example, in many cases (Dirichlet boundary conditions around a region R containing only regular internal points) it may be shown that the maximum error is bounded in modulus by $M_4 L^2 \Delta^2/24$, where L is the radius of a circle enclosing the entire region and M_4 is an upper bound on the modulus of the fourth partial derivative, with respect to either direction, of the solution, taken over all the points. Since M_4 is not known until the solution itself is known, this estimate cannot be made *a priori*; but one can sometimes estimate $\partial^4 T/\partial x^2 \partial y^2$ by means of the difference quotient

$$\frac{\partial^4 T}{\partial x^2 \partial y^2} = \frac{1}{\Delta^2}[(T_{NE} - 2T_N + T_{NW}) - 2(T_E - 2T_P + T_W) + (T_{SE} - 2T_S + T_{SW})] \quad (4.4.4)$$

using computed values, and take this as an estimate of M_4, noting that

$$\frac{\partial^4 T}{\partial x^2 \partial y^2} = -\frac{\partial^4 T}{\partial x^4} = -\frac{\partial^4 T}{\partial y^4}$$

by virtue of the basic 2-D Laplace equation. In eqn. (4.4.4) obvious compass notation is used.

The latter type of error is committed by using the solution of the FDEs instead of the PDE, incurring a discretisation (or truncation) error at every point of the grid. One normally expects this error to reduce as the grid is refined, and one can take advantage of this property as follows. For square meshes of size Δ, the errors for the five- and nine-point formulae, eqns. (3.3.7) and (3.3.8) of Chapter 3, are bounded by quantities which are proportional to Δ^2 and Δ^6 respectively. In proceeding from a coarse to a finer mesh, suppose T_1 and T_2 are the calculated solution values of the temperature at a certain fixed spatial location, obtained through using two different mesh sizes, Δ_1 and Δ_2 respectively, where $\Delta_1 > \Delta_2$. Then a better approximation than either T_1 or T_2 at this point is (see Bayley *et al.*, 1972; Razelos, 1973):

$$\text{Five-point formula:} \quad T = \frac{T_2 - \left(\dfrac{\Delta_2}{\Delta_1}\right)^2 T_1}{1 - \left(\dfrac{\Delta_2}{\Delta_1}\right)^2}$$

$$\text{Nine-point formula:} \quad T = \frac{T_2 - \left(\dfrac{\Delta_2}{\Delta_1}\right)^6 T_1}{1 - \left(\dfrac{\Delta_2}{\Delta_1}\right)^6} \quad (4.4.5)$$

Often, on refining a mesh, one just halves the grid spacings so that the case $\Delta_1 = 2\Delta_2$ is especially important. In this case eqn. (4.4.5) becomes

$$\text{Five-point formula:} \quad T = \frac{(4T_2 - T_1)}{3}$$

$$\text{Nine-point formula:} \quad T = \frac{(64T_2 - T_1)}{63} \tag{4.4.6}$$

4.4.3. Example

A square plate, shown in Fig. 4.21, has one face at 150 °C and the other three at 50 °C. Temperatures are computed with the five-point formula

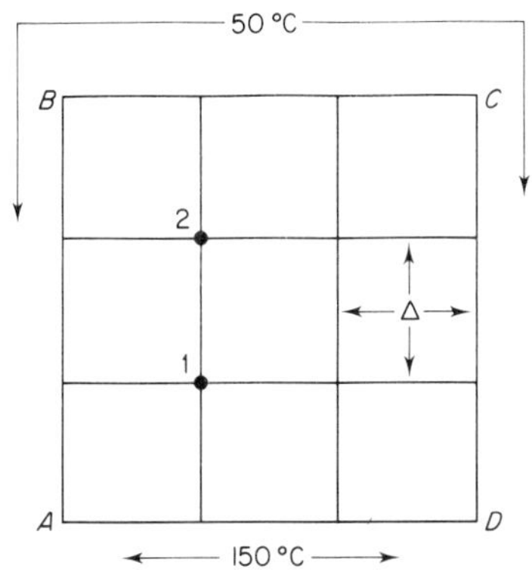

FIG. 4.21. Rectangular plate with base side held at 150 °C and other faces held at 50 °C.

using two grids, one with mesh size refined to Δ/2. The computed temperatures at two fixed spatial positions labelled 1 and 2 were found, then extrapolated using eqn. (4.4.6) and compared with the solution obtained with the nine-point formula on the coarser grid, and the exact solution obtained by separation of variables and Fourier series at these locations (Carslaw & Jaeger, 1959). The results are shown in Table 4.4. In using the nine-point formula, temperatures at the corners *A* and *D* are required, and these are overspecified in the problem: average values of 100 °C are used.

Notice the improvement obtained by halving the mesh size with the five-point formula or using the nine-point formula. The extrapolated temperatures improve the accuracy by about a factor of two.

In extrapolating to the limit of zero mesh size, eqns. (4.4.5) and (4.4.6)

TABLE 4.4
COMPARISON OF TEMPERATURES DETERMINED BY DIFFERENT METHODS

Method	T_1	T_2
Five-point, mesh size Δ	87·50	62·50
Five-point, mesh size $\Delta/2$	87·88	62·12
Extrapolation, using eqn. (4.4.6)	88·01	61·99
Nine-point, mesh size Δ	88·10	61·90
Exact analytic solution	88·18	61·93

required knowledge of the order of accuracy of the finite difference representation. If this is not known, an extrapolation formula can still be obtained but it is more complicated. Suppose as before that T_1 and T_2 are finite difference solutions obtained at the same point with mesh sizes Δ_1 and Δ_2 respectively. Taking the leading error term to be proportional to Δ^n, then

$$T - T_1 = A\Delta_1^n$$

$$T - T_2 = A\Delta_2^n$$

where A and n are unknown constants. Eliminating A yields

$$T = \frac{\Delta_2^n T_1 - \Delta_1^n T_2}{\Delta_2^n - \Delta_1^n} \tag{4.4.7}$$

Substituting $n = 2$ or 6 along with $\Delta_1 = 2\Delta_2$ gives eqns.(4.4.5) and (4.4.6). When n is not known, however, it can be estimated by finding another finite difference solution T_3 at the same point with a mesh size Δ_3 where $\Delta_1 = 2\Delta_2 = 4\Delta_3$. Then n is estimated from

$$2^n = \frac{T_2 - T_1}{T_3 - T_2} \tag{4.4.8}$$

4.4.4. Deferred-Correction Method

In this method, one must calculate an initial approximate solution, difference it to obtain correction terms, add these into the FDEs and keep as constants, and then solve the amended FDEs which then yield a more accurate solution.

Consider Laplace's equation in a 2-D cartesian system, with a uniform rectangular grid. Section 3.1 of Chapter 3 outlined the way in which various FDEs approximate to derivatives. In particular, it can be shown that

$$\left.\frac{d^2 f}{dx^2}\right|_P = \frac{1}{(\Delta x)^2}\left[\delta^2 - \frac{1}{12}\delta^4 + \frac{1}{90}\delta^6 - \cdots\right] f_0$$

This is usually truncated after the first term to obtain the familiar

$$\left.\frac{\mathrm{d}^2 f}{\mathrm{d}x^2}\right|_P = \frac{f_E - 2f_P + f_W}{(\Delta x)^2}$$

in the usual compass notation. This, combined with the second-order y-derivative, gives the usual five-point formula with accuracy order $(\Delta x)^2$. Including the second term gives

$$\left.\frac{\mathrm{d}^2 f}{\mathrm{d}x^2}\right|_P = \frac{f_E - 2f_P + f_W}{(\Delta x)^2} + \frac{1}{(\Delta x)^2}[f_{EE} - 4f_E + 6f_P - 4f_W + f_{WW}]$$

and similarly for the y-derivative. Rather than substitute all these expressions into

$$\frac{\partial^2 T}{\partial x^2} + \frac{\partial^2 T}{\partial y^2} = 0$$

which would give a large star formula with accuracy order $(\Delta x)^4 + (\Delta y)^4$ using nine points, and be tricky to handle near boundaries, the correction part is calculated once and for all based on the initial approximate solution and subsequently retained as a constant. Iteration now, with this correction to the usual five-point formula, still leads to an improved solution.

4.5. STABILITY AND CONVERGENCE*

4.5.1. Errors

Again, as in the previous section, the error refers to the difference between the approximate calculated solution of the FDEs and the exact solution of the governing PDE, although one may not know and there may be no formal expression for this exact solution. There are two types of errors. The first is called the truncation error, due to computing with derivatives replaced by finite differences, and this depends on the initial given temperature distribution, the boundary conditions, the choice of finite difference scheme and the choice of the Fourier number Fo used in the computation. Convergence is the expression used to indicate that refining the space–time grid makes the approximate computed solution approach the exact solution. Not all finite difference simulations of PDEs possess this convergence property. The second type of errors are numerical: round-off errors which are caused by the finite significant figure restriction used in any

* This section is concerned with transient problems, and follows the work of Razelos (1973).

calculation, and errors which, for a given space–time grid associated with the Fourier number Fo, are inherent in certain finite difference schemes being unstable with regard to the propagation and growth of oscillatory values as the computed solution. The following illustration exemplifies this.

4.5.2. Example

Figure 4.22 shows an infinite flat plate of thickness 0·25 m having thermal diffusivity $\alpha = 3{\cdot}472 \times 10^{-6}\,\mathrm{m^2/s}$, and initially at temperature 0 °C. At time

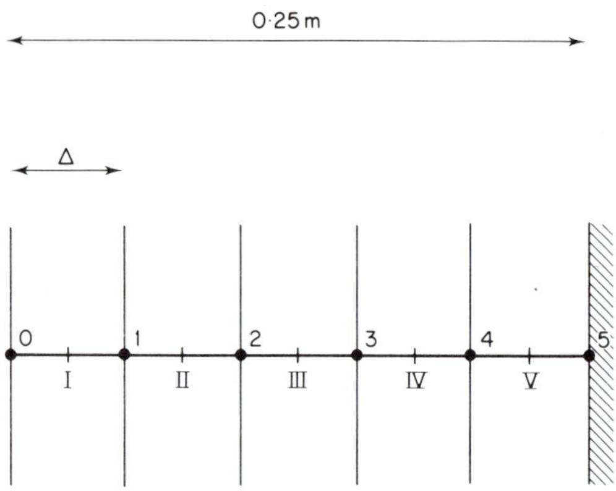

FIG. 4.22. Subdivision of 1-D plate for transient numerical computation.

$t = 0$, one face of the plate is suddenly increased to, and maintained at, 100 °C (which is equivalent to taking $T = 100$ and $h = \infty$ with convective boundary conditions) while the other is insulated, and 1-D transient conduction takes place. Taking five equal sections as shown and using an explicit FDE (eqn. (3.5.5) of Chapter 3) at each point labelled 1 to 4, one can compute the time–temperature history at each of the points. Tables 4.5 and 4.6 show the results of calculations with Fourier number Fo = 1 and 0·25 respectively, for which the corresponding time increments, obtained from $\Delta t = \mathrm{Fo}\,(\Delta x)^2/\alpha$, are 12 min and 3 min respectively.

4.5.3. Stability

Inspection of Table 4.5 reveals that the computed values oscillate with increasing magnitude as the calculation proceeds. This instability phenomenon, obviously yielding an unsatisfactory computed solution, has nothing to do with round-off errors, but is a property of the FDE, initial and boundary conditions. (However, rounding errors tend to grow with time in unstable schemes, but damp out in stable schemes, as the forward march in

TABLE 4.5
FLAT PLATE TEMPERATURE RESPONSE WITH FOURIER NUMBER Fo = 1

t		Nodal point temperature (°C)				
(*min*)	0	1	2	3	4	5
0	100	0	0	0	0	0
12	100	100	0	0	0	0
24	100	0	100	0	0	0
36	100	200	−100	100	0	0
48	100	−200	400	−200	100	0
60	100	700	−800	700	−300	
72	100	−1400	2200	−1800	1200	

time progresses: observe the accuracy shown in Table 4.6, despite the presence of round-off errors which do not occur in Table 4.5.)

Evidently the space and time increments Δx and Δt cannot be selected arbitrarily, but they must satisfy certain stability criteria which may be established analytically. However, simple stability criteria may be developed which are adequate for practical applications of explicit schemes, based on an elementary argument. Consider the general explicit transient

TABLE 4.6
FLAT PLATE TEMPERATURE RESPONSE WITH FOURIER NUMBER Fo = 0·25

t		Nodal point temperature (°C)				
(*min*)	0	1	2	3	4	5
0	100	0	0	0	0	0
3	100	25·000 0	0	0	0	0
6	100	37·500 0	6·250 0	0	0	0
9	100	45·312 5	12·500 0	1·562 5	0	0
12	100	50·781 2	17·968 7	3·906 3	0·390 6	0
15	100	54·882 8	22·656 2	6·543 0	1·171 9	0·195 3
18	100	58·105 4	26·684 5	9·228 5	2·270 5	0·683 6
21	100	60·723 8	30·175 7	11·853 0	3·613 3	1·477 1
24	100	62·905 8	33·232 1	14·373 7	5·139 2	2·545 2
27	100	64·760 9	35·935 9	16·779 7	6·799 3	3·842 2
30	100	66·364 4	38·353 1	19·073 6	8·555 1	5·320 7
Exact solution (Carslaw & Jaeger, 1959) 30 min	100	65·178 7	37·183 0	18·012 4	8·289 7	5·496 3

FDE (eqn. (3.5.3) of Chapter 3):

$$T_O + T_I + T_E + T_W + T_N + T_S - \left(6 - \frac{1}{\text{Fo}}\right)T_P = \frac{1}{\text{Fo}}T'_P \quad (4.5.1)$$

where the Fourier number $\text{Fo} = \alpha\Delta t/\Delta^2 = k\Delta t/\rho C_p\Delta^2$. The coefficients of the T_j's ($j = O, I, N, S, E$ and W), the temperatures at all the six nodes surrounding the node P, are all positive and equal to 1, in fact, for this case of a uniform regular mesh with $\Delta = \Delta x = \Delta y = \Delta z$. If the coefficient of T_P, the temperature at P at time-level t, were negative, then this formula would dictate that the lower T_P was, the higher T'_P would be at the new time-level $t + \Delta t$. But this would violate thermodynamic principles and it follows that a necessary condition for non-violation is that Δt should be chosen, for a given spatial subdivision, so that the coefficient of T_P is non-negative. This turns out to be in fact a sufficient condition for stability for explicit FDEs. Applying this simple rule to various explicit FDEs, one may deduce the following restrictions on the Fourier number $\text{Fo} = k\Delta t/\rho C_p\Delta^2$ and the Biot number $\text{Bi} = h\Delta/k$:

Interior nodes:

1-D cartesian	$\text{Fo} \leq \frac{1}{2}$	
2-D cartesian	$\text{Fo} \leq \frac{1}{4}$	
3-D cartesian	$\text{Fo} \leq \frac{1}{6}$	
Hexagonal	$\text{Fo} \leq \frac{3}{4}$	(4.5.2)

Boundary nodes (convective):

1-D cartesian	$\text{Fo} \leq 1/[2(1 + \text{Bi})]$	
1-D cartesian (displaced grid)	$\text{Fo} \leq (2 + \text{Bi})/(2 + 3\,\text{Bi})$	
2-D cartesian	$\text{Fo} \leq 1/[2(2 + \text{Bi})]$	
2-D cartesian (corner node)	$\text{Fo} \leq 1/[4(1 + \text{Bi})]$	
3-D cartesian	$\text{Fo} \leq 1/[2(3 + \text{Bi})]$	(4.5.3)

According to this simple rule, stability should not be affected by the addition of a constant internal heat generation term.

Obviously, conditions of eqn. (4.5.3) without a displaced grid system are unfavourable for systems with a large value of Bi, that is, a large ratio of surface conductance to internal conductance. For example, given h, k and Δ in a 1-D transient problem to be solved with an explicit technique, the time-step Δt is restricted via

$$\Delta t = \frac{\Delta^2}{\alpha}\text{Fo} \leq \frac{\Delta^2}{2\alpha(1 + \text{Bi})} \quad (4.5.4)$$

Thus when Bi is large, Δt may have to be taken unnecessarily small for stability reasons, resulting in a substantial increase in labour without specially increasing the accuracy. A simple, less accurate way of relieving this problem without resorting to an implicit method is to relocate the nodes half a grid interval sideways as shown by nodes I to V in Fig. 4.22. The resulting stability criterion for the convective boundary is more relaxed, as indicated by the second of eqn. (4.5.3) as opposed to the first.

Stability criteria may also be developed for the weighted-average implicit formulae, eqns. (3.5.12) to (3.5.14) of Chapter 3, though not quite so simply as for the explicit forms just considered. In particular, the six-point formula, eqn. (3.5.13) of Chapter 3, is stable at interior points provided

$$\mathrm{Fo} \leq \frac{1}{2(1 - 2\lambda)} \tag{4.5.5}$$

where λ is the implicit weighting factor, viz. $\lambda = 0$ gives the explicit scheme, $\lambda = 1$ the fully implicit scheme. Notice that if $\lambda \geq 0{\cdot}5$ (*i.e.* $0{\cdot}5 \leq \lambda \leq 1$), then the FDE simulation is always stable. This is a general result which carries over to the fully implicit scheme ($\lambda = 1$) which is always stable.

If T_P is involved in the heat generation expression, the analogous expression to eqn. (4.5.1) must be obtained.

4.5.4. Convergence

Sometimes, refining the time-step Δt and distance-steps Δx, for example, in the solution of a transient problem generates a finite difference representation which is stable, but the solution of which gives values which tend to the solution of a different PDE. If this is not the case, the finite difference representation is said to be consistent with the given PDE. It can be shown that many systems which are both stable and consistent are automatically convergent (a desirable property), and then refining the grid improves the accuracy of solution. With the explicit method for a transient problem in one space dimension, the $\Delta t/(\Delta x)^2$ restriction demands excessive computation in order to generate stable answers. Although the solution values then calculated are extremely accurate, they may be far more accurate than required, and very expensive in computer time to acquire. A less, but sufficiently, accurate solution may be obtained by using an implicit method on a coarser space and time grid.

4.5.5. Accuracy

Concerning accuracy:

(i) Round-off errors have little or no effect if the system is stable.
(ii) Truncation error has the most prominent effect on accuracy. At

interior nodes the error of the generalised six-point formula is

$$|E_t| \leq \frac{\alpha^2 \Delta t}{2} \cdot \frac{\partial^4 T}{\partial x^4} \cdot \left(1 - 2\lambda - \frac{1}{6\,\mathrm{Fo}}\right) + A(\Delta t)^2 + B(\Delta x)^4 \qquad (4.5.6)$$

where E_t is the truncation error, T represents the exact solution and A and B are constants. For degenerate versions of this:

Explicit: $|E_t| \leq A_1 \Delta t + B_1(\Delta x)^2$

Implicit: $|E_t| \leq A_2 \Delta t + B_2(\Delta x)^2$

Mid-difference (Crank–Nicolson): $|E_t| \leq A_3 (\Delta t)^2 + B_3(\Delta x)^2$ (4.5.7)

If $0 \leq \lambda < \frac{1}{2}$, higher accuracy of the generalised six-point formula can be obtained by setting

$$\mathrm{Fo} = \frac{1}{6(1 - 2\lambda)} \qquad (4.5.8)$$

in which case the first term on the right of eqn. (4.5.6) vanishes. For example, if $\lambda = 0$ (explicit representation), then taking $\mathrm{Fo} = \frac{1}{6}$ is best for reducing truncation error. Just as complications with boundary conditions affected the stability criterion adversely, so it is with the introduction of errors. Thus if sudden changes occur at the boundaries, one could begin calculations with a small time-step (say with $\mathrm{Fo} = \frac{1}{6}$) and after some time, as the sudden change in the near boundary temperature field has dampened down, lengthen it to correspond to $\mathrm{Fo} = \frac{1}{4}$ or $\frac{1}{2}$, for example.

BIBLIOGRAPHY

AMES, W. F. (1969). *Numerical Methods for Partial Differential Equations*. Nelson, London.

BAYLEY, F. J., OWEN, J. M. & TURNER, A. B. (1972). *Heat Transfer*. Nelson, London.

CARSLAW, H. S. & JAEGER, J. (1959). *Conduction of Heat in Solids*. Clarendon Press, Oxford.

DUSINBERRE, G. M. (1961). *Heat Transfer Calculations by Finite Differences*. International Textbook Co., Scranton, Pa.

GEBHART, B. (1971). *Heat Transfer*, 2nd ed. McGraw-Hill, New York.

RAZELOS, P. (1973). Methods of obtaining approximate solutions. Section 4, pp. 4-1 to 4-78, in *Handbook of Heat Transfer*, ed. W. M. Rohsenow and J. P. Hartnett. McGraw-Hill, New York.

SCHNEIDER, P. J. (1973). Conduction. Section 3, pp. 3–1 to 3–134, in *Handbook of Heat Transfer*, ed. W. M. Rohsenow and J. P. Hartnett. McGraw-Hill, New York.

SMITH, G. D. (1965). *Numerical Solution of Partial Differential Equations*. Oxford University Press, London.

WALSH, J. (1966). Elliptic and parabolic equations. Chapter 5, pp. 99–116, in *Numerical Analysis: An Introduction*, ed. J. Walsh. Academic Press, London.

YOUNG, D. (1962). The numerical solution of elliptic and parabolic partial differential equations. In: *A Survey of Numerical Analysis*, ed. J. Todd. McGraw-Hill, New York.

CHAPTER 5

Application of Finite Difference Equations to Simple Steady-State Problems*

5.1. 1-D STEADY-STATE PROBLEMS: DIRECT TRIDIAGONAL MATRIX ALGORITHM SOLUTION

5.1.1. Simple Problem

Consider the 1-D problem shown in Fig. 5.1 of a lagged square-sectioned $10 \times 10\,\text{cm}^2$ metal bar of length 50 cm carrying an electric current and so generating heat internally. For simplicity it is assumed that conditions across any cross-section of the bar are constant and that the ends are maintained at 10 °C. Taking four interior node points, labelled 1 to 4 in the figure, gives the spacing

$$\Delta x = \frac{50}{(4+1)} = 10\,\text{cm}$$

A typical cell control volume for node 2 is also shown, and since the bar is lagged, the temperature T_2 is influenced by T_1 and T_3 only by way of their conductances K_{12} and K_{32}. Since the cross-sectional area A is $0{\cdot}1 \times 0{\cdot}1 = 0{\cdot}01\,\text{m}^2$, these conductances are

$$\begin{aligned} K_{12} &= \frac{kA}{\Delta x} \\ &= 50 \times 0{\cdot}01/0{\cdot}1 \\ &= 5 \end{aligned}$$

and exactly the same for K_{32}. Other conductances are similarly the same value of 5. Here we have taken the thermal conductivity of the bar k to be

* This chapter draws from the previous chapters and the bibliography listed at the end, which provides further reading.

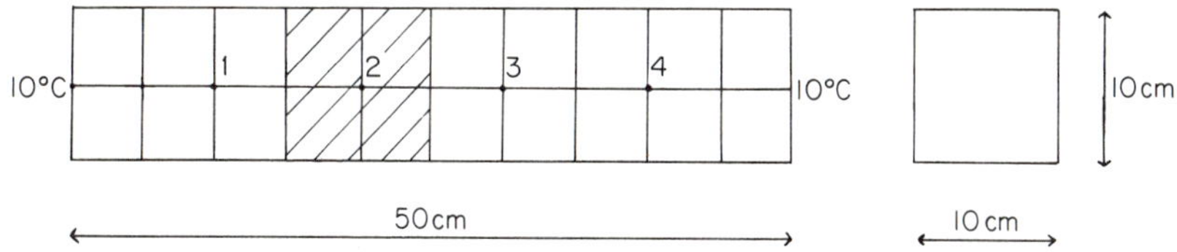

FIG. 5.1. Square-sectioned metal bar.

50 W/mK; we also assume that the internal heat generation H is 4×10^4 W/m^3. The cell control volume associated with node 2 has volume $V = 0{\cdot}1^3 \text{m}^3 = 10^{-3} \text{m}^3$.

To derive the FDE for node 2, making use of the cell control volume shaded in the figure, one simply appeals to eqn. (3.4.2) of Chapter 3:

$$\sum_j K_{2j}(T_j - T_2) + HV_2 = 0 \tag{5.1.1}$$

where the summation is over neighbouring nodes, $j = 1$ and 3 only. Hence

$$k\frac{A}{\Delta x}(T_1 - T_2) + k\frac{A}{\Delta x}(T_3 - T_2) + H \cdot \Delta x \cdot A = 0$$

which leads to

$$T_1 - 2T_2 + T_3 + 8 = 0 \tag{5.1.2}$$

This equation could equally well have been obtained by direct finite difference approximation to the governing 1-D cartesian steady-state PDE:

$$k\frac{\partial^2 T}{\partial x^2} + H = 0 \tag{5.1.3}$$

The appropriate FDE is

$$k\frac{T_1 - 2T_2 + T_3}{(\Delta x)^2} + H = 0 \tag{5.1.4}$$

from which eqn. (5.1.2) follows at once. Equation (5.1.4) is precisely eqn. (3.3.3) of Chapter 3, which could equally well have been written down as the basic equation.

Writing down the appropriate equation for each of the other interior nodes, and incorporating the end-point temperatures $T_0 = T_5 = 10\,°\text{C}$, gives the following equation set for T_1, T_2, T_3 and T_4 (the temperatures at the four points along the bar):

$$\begin{aligned}
-2T_1 + T_2 \qquad\qquad &= -8 - 10 = -18 \\
T_1 - 2T_2 + T_3 \qquad &= -8 \\
T_2 - 2T_3 + T_4 &= -8 \\
T_3 - 2T_4 &= -8 - 10 = -18
\end{aligned} \tag{5.1.5}$$

This is what is known as a tridiagonal equation set—each equation has exactly three unknowns in a particular order except the first and last equations which have exactly two unknowns. Direct elimination methods may be applied to obtain the solution. In particular, direct Gauss elimination applied to such a set of equations takes a particularly simple form and the technique is known as the tridiagonal matrix algorithm (TDMA). Regardless of the number of interior nodes taken, the form of the equations is still tridiagonal and the TDMA is readily applicable in order to obtain the solution. The technique is further discussed and illustrated with Program 2.2 in Section 7.2 of Chapter 7. It is generally not worthwhile to use an iteration procedure unless more unknowns appear in each equation; then the techniques to be described in Sections 5.2 and 5.3 of this chapter will be needed.

5.2. 2-D STEADY-STATE PROBLEMS: ITERATIVE SOLUTION

5.2.1. Simple Problem

Two-dimensional steady-state heat flow problems are governed by the PDE

$$k\left(\frac{\partial^2 T}{\partial x^2} + \frac{\partial^2 T}{\partial y^2}\right) + H = 0 \tag{5.2.1}$$

and how to replace this by a set of FDEs has been considered earlier. Using a square mesh in a cartesian coordinate system, consider for simplicity the problem of Fig. 5.2. Here a rectangular region with sides in the proportion 1:2 has been overlaid with a square grid system resulting in three interior nodes labelled 1, 2 and 3 at which the temperature is unknown. Assume further that all boundary temperatures at nodes $A, B, \ldots, H$ are known (Dirichlet boundary conditions) and that there is no internal heat

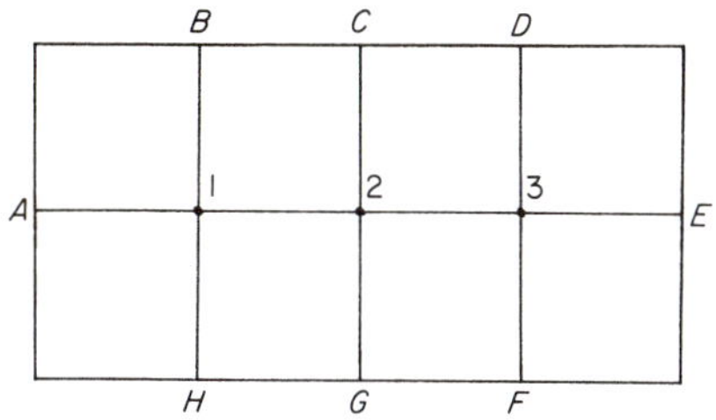

FIG. 5.2. 2-D steady-state problem with a coarse mesh.

generation. Direct replacement of eqn. (5.2.1) or energy balance techniques show that the appropriate FDE is (see eqn. (3.3.2) of Chapter 3):

$$T_N + T_S + T_E + T_W - 4T_P = 0 \tag{5.2.2}$$

using obvious compass notation. Hence in order to calculate the three temperatures T_1, T_2 and T_3, one must apply eqn. (5.2.2) to each point 1, 2 and 3 in turn, incorporate the boundary temperatures and then solve the resulting set of three simultaneous linear algebraic equations. Applying eqn. (5.2.2) to each node in turn:

Node 1: $T_B + T_H + T_2 + T_A - 4T_1 = 0$

Node 2: $T_C + T_G + T_3 + T_1 - 4T_2 = 0$

Node 3: $T_D + T_F + T_E + T_2 - 4T_3 = 0$ (5.2.3)

Incorporating the boundary conditions, which are the fixed boundary temperatures detailed in Fig. 5.3, yields the following set of three equations in the three unknowns T_1, T_2 and T_3:

$$\begin{aligned} 4T_1 - T_2 \phantom{{}+4T_3} &= 400 \\ -T_1 + 4T_2 - T_3 &= 400 \\ -T_2 + 4T_3 &= 1360 \end{aligned} \tag{5.2.4}$$

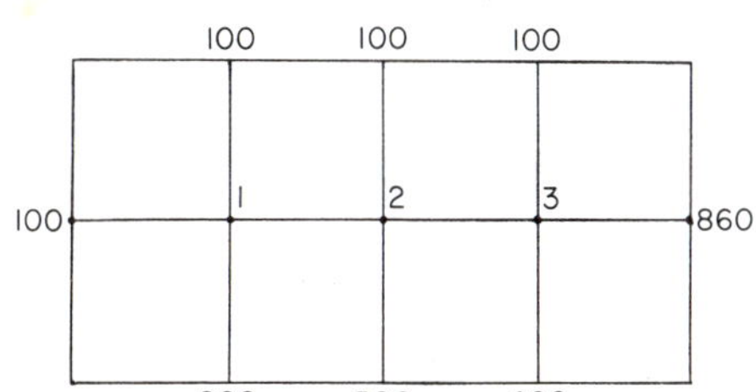

FIG. 5.3. Boundary conditions for the 2-D steady-state problem (temperatures in °C).

These equations can be solved quickly by any direct elimination technique (since there are merely three equations in three unknowns) to give the solution:

$$\begin{aligned} T_1 &= 160\,^\circ\text{C} \\ T_2 &= 240\,^\circ\text{C} \\ T_3 &= 400\,^\circ\text{C} \end{aligned} \tag{5.2.5}$$

Clearly, however, with a fine mesh system and many internal nodes the resulting set of equations will be large and other solution techniques will be required (see Section 5.3 and Chapter 7).

Equation (5.2.5) is the exact solution of eqn. (5.2.4), which represents a coarse grid finite difference simulation of the actual problem as determined by the governing PDE (eqn. (5.2.1)) and the boundary conditions shown in Fig. 5.3. As discussed in Chapter 4, Section 4.4, there is little hope that this corresponds at these three points to the exact analytic solution of eqn. (5.2.1) subject to the appropriate boundary conditions—we are solving FDEs which possess a discretisation error. However, in this section we are concerned with illustrating the use of iterative techniques which are capable of only approximate solution of the appropriate FDEs like eqn. (5.2.4), though one can often iterate so as to solve them as accurately as desired.

The coarse mesh for this problem has been chosen deliberately so as to illustrate some of the iterative methods of solution; this problem is returned to in Chapter 7 where simulation and solution with a finer mesh are described. Three methods of solving these three FDEs for the temperatures at nodes 1, 2 and 3 will now be outlined; the opportunity is also taken to introduce the idea of solution via a Fortran 4 computer program. Clearly this is not necessary for this simple problem, but the build-up of ideas, which are carried through to more complex problems in later chapters, will be particularly useful.

The basis of all iterative methods is that an approximate solution is guessed for each point at which solution is sought (it is often as vague as $T_1 = T_2 = T_3 = 0$) and the T values are then repeatedly updated, usually sequentially, many times until repeated updating results in very little change to the T values. We are then said to have converged on to the solution of the equations, which is taken as the latest available set of T values. Notice that when T_1, T_2 and T_3 are solutions of eqn. (5.2.4), then the residuals

$$\begin{aligned} R_1 &= 400 - 4T_1 + T_2 \\ R_2 &= 400 + T_1 - 4T_2 + T_3 \\ R_3 &= 1360 + T_2 - 4T_3 \end{aligned} \tag{5.2.6}$$

are each equal to zero, for they measure the departure from exactness of the solution of the FDE at the point in question. With initial guesses for the T's, in general the R's are non-zero and it is the aim with all iterative methods to update the T's repeatedly by some technique, so as to reduce them as near as possible to zero. Updating techniques differ in complexity and efficiency in their route to the solution. Three methods are now to be described in the context of the simple problem. In Section 5.3 relaxation methods are described in a more general context.

5.2.2. Jacobi Iteration

Initial approximate values are given to T_1, T_2 and T_3, these being guesses to the required solution. At this first, and at any later, iteration stage the residuals R_1, R_2 and R_3 are calculated. At each stage, in order to make a further iteration, each T_i $(1 \leq i \leq 3)$ is updated via

$$T_i' = T_i + \frac{R_i}{4} \qquad (1 \leq i \leq 3) \tag{5.2.7}$$

where the prime is used to denote an updated value based on previous iterate non-primed values of T. Thus

$$\begin{matrix}\text{Updated value} \\ \text{of } T \text{ at node } i\end{matrix} = \begin{matrix}\text{Previous value} \\ \text{of } T \text{ at node } i\end{matrix} + \begin{matrix}\text{Residual/4} \\ \text{at node } i\end{matrix}$$

This has the effect of temporarily reducing the residual R_i to zero at node i—temporarily because updating an adjacent node immediately destroys this. Updating proceeds sequentially—T_1, T_2 and T_3 for one iteration. New temperatures are not used at once, but only after each iteration, one iteration being the updating of all the temperatures in the equation set. Convergence results after several iterations; the method is simple to apply but not very efficient in computational effort required to achieve a certain accuracy. Observe that the update of, for example, the temperature at node 1 is equivalent to

$$\begin{aligned} T_1' &= T_1 + \frac{(400 - 4T_1 + T_2)}{4} \\ &= \frac{(400 + T_2)}{4} \\ &= \frac{(T_A + T_B + T_H + T_2)}{4} \end{aligned} \tag{5.2.8}$$

which could have been obtained from eqn. (5.2.3) by simple rearrangement to isolate the temperature at node 1, T_1, on the left and transfer the other terms to the right-hand side. Iteration then proceeds by updating T_1 to T_1' based on previous iterate values on the right. It is important to note also that the coefficient of T_1 in eqn. (5.2.3) is the largest of the terms on the left; the T_1 term is dominant in the equation. This is defined precisely later in Section 7.2 of Chapter 7, but for now it is enough to notice that the absolute value of the coefficient of T_1 is at least as large as the sum of the absolute values of the coefficients of the other unknowns. Similar comments apply also to the equations for the other two nodes. In practice, when an FDE is written down for a particular node, the unknown at that node is generally dominant in its equation. This dominance is a requirement for the convergence of these iterative techniques and is discussed more fully later in Section 7.2 of Chapter 7.

Returning to the current problem, the updating is

$$\begin{aligned} T_1' &= T_1 + \frac{R_1}{4} \\ T_2' &= T_2 + \frac{R_2}{4} \\ T_3' &= T_3 + \frac{R_3}{4} \end{aligned} \qquad (5.2.9)$$

or

$$\begin{aligned} T_1' &= T_1 + \frac{(T_A + T_B + T_2 + T_H)}{4} \\ T_2' &= T_2 + \frac{(T_1 + T_C + T_3 + T_G)}{4} \\ T_3' &= T_3 + \frac{(T_2 + T_D + T_E + T_F)}{4} \end{aligned} \qquad (5.2.10)$$

which, on substituting for the R's, is simply

$$\begin{aligned} T_1' &= (400 + T_2)/4 \\ T_2' &= (400 + T_1 + T_3)/4 \\ T_3' &= (1360 + T_2)/4 \end{aligned} \qquad (5.2.11)$$

or, illuminating a little,

$$\begin{aligned} T_1' &= (100 + 200 + 100 + T_2)/4 \\ T_2' &= (100 + 300 + T_1 + T_3)/4 \\ T_3' &= (100 + 400 + 860 + T_2)/4 \end{aligned} \qquad (5.2.12)$$

or, in terms of the original notation of Fig. 5.2,

$$\begin{aligned} T_1' &= (T_B + T_H + T_A \qquad + T_2 \qquad)/4 \\ T_2' &= (T_C + T_G \qquad + T_1 \qquad + T_3)/4 \\ T_3' &= (T_D + T_F + T_E \qquad + T_2 \qquad)/4 \end{aligned} \tag{5.2.13}$$

This last set of equations could have been obtained directly by rearrangement of eqn. (5.2.3).

Sometimes it is convenient to work with the residual after an equation has been normalised to obtain the unit coefficient of its dominant term. In this case, the residual r_i is related to R_i (at any node point i) by

$$r_i = \frac{R_i}{4} \qquad (1 \leq i \leq 3) \tag{5.2.14}$$

and the update formula, eqn. (5.2.7), becomes

$$T_i' = T_i + r_i \tag{5.2.15}$$

Working with the R_i's, Table 5.1 illustrates the calculation, beginning with zero estimates, of seven iterations in the search for the solution of eqn. (5.2.4) using eqn. (5.2.7). Notice that whenever eqn. (5.2.7) is applied to node i, its residual is temporarily reduced to zero (which is not recorded) and the residual at an adjacent point is altered. In fact if T_i is altered by amount ΔT_i, R_i changes by amount $-4\Delta T_i$ and the R_j's of adjacent points change by amount $+\Delta T_i$. In Table 5.1 an entire row of new T_1, T_2 and T_3 are calculated, then the three new residuals R_1, R_2 and R_3 are calculated. After seven iterations, the approximate solution is

$$\begin{aligned} T_1 &= 159{\cdot}7 \quad (160) \\ T_2 &= 239{\cdot}7 \quad (240) \\ T_3 &= 399{\cdot}9 \quad (400) \end{aligned} \tag{5.2.16}$$

TABLE 5.1

JACOBI ITERATION FOR THE PROBLEM OF FIG. 5.2

Iteration number	T_1	R_1	T_2	R_2	T_3	R_3
0	0	400	0	400	0	1 360
1	100	100	100	440	340	100
2	125	110	210	50	365	110
3	152·5	12·5	222·5	55	392·5	12·5
4	155·6	13·9	236·3	6	395·6	13·9
5	159·1	1·4	237·8	7	399·1	1·4
6	159·45	1·0	239·6	0·5	399·45	1·8
7	159·7		239·7		399·9	

with the exact solution shown in parentheses for comparison. More iterations will take us even nearer to this exact solution.

Rather than update the T_i's, then calculate the R_i's at each iteration, one could merely calculate the sequence of T's without regard to the R's. In this case one uses eqn. (5.2.11) directly instead of eqn. (5.2.9) and less calculation is required. Obviously the same sequence of T's results, but one's knowledge of how near one is to the solution is limited. It need not be, for in every case (from eqns. (5.2.7) or (5.2.9))

$$R_i = 4(T_i' - T_i) \tag{5.2.17}$$

so that the previous residual can easily be calculated from the current and previous temperatures at the node in question.

PROGRAM 2.1 (see p. 250)

A specific computer program, set to performing 20 iterations of eqn. (5.2.12), is listed as Program 2.1 and serves as an introduction to the idea of solution of problems via a Fortran 4 computer program. Very little in the way of programming maturity is required for its comprehension, and the comment cards adequately describe the steps in the code.

PROGRAM 2.2 (see p. 251)

Program 2.2 performs the same task in a more general manner: the boundary temperatures are now allocated storage space and filled via a DATA statement and the heart of the program, the iteration loop update calculation, utilises the more general equations, eqn. (5.2.13). Still set just to 20 iterations, the listing includes the output of headings, initial and intermediate output.

PROGRAM 2.3 (see p. 252)

Program 2.3, on the other hand, introduces a 2-D array as shown in Fig. 5.4 to cover the points at which temperature is known (on the boundary) or required to be found (the three interior nodes). Although updates only on the three points (2, 2), (3, 2) and (4, 2) are required, two nested DO loops are included as a reminder that, with a finer mesh and a rectangular array of points to update, it is only necessary to make suitable changes to the range of these loops. The update statement for any of these points takes the form

$$\mathrm{TP(I,J) = [T(I-1,J) + T(I+1,J) + T(I,J-1) + T(I,J+1)]/4{\cdot}0}$$

as shown in the DO 100 loop. The program also allows the number of i and j grid lines to be specified as IMAX and JMAX, and the major DO loops range over I = 2, IMM1 and J = 2, JMM1 where IMM1 = IMAX − 1 and JMM1 = JMAX − 1. The iteration loop L = 1, LMAX continues until aborted, this being so when the maximum fraction change in any T, FMAX, or the maximum absolute sum of the residuals, RES, becomes smaller than a specified value, taken as 0·005.

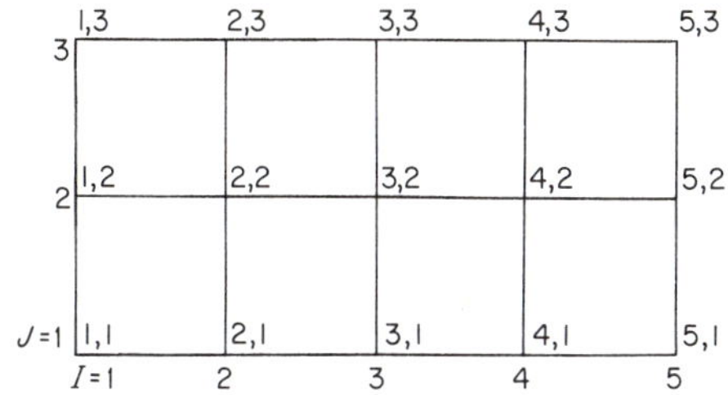

FIG. 5.4. The 2-D (I, J) array and notation for the problem of Fig. 5.2.

5.2.3. Gauss–Seidel Iteration

The Gauss–Seidel iteration method is similar to the Jacobi iteration method. The sole difference is that, when sequentially iterating with the update equations, new updated values are used as soon as they are available. The governing equation set, eqns. (5.2.3) and (5.2.4), is exactly the same as before. On rearranging for the dominant diagonal terms, one takes eqn. (5.2.11) or, equivalently, eqn. (5.2.13) in the form

$$\begin{aligned} T_1' &= (400 \qquad + T_2 \qquad)/4 \\ T_2' &= (400 + T_1' \qquad + T_3)/4 \\ T_3' &= (1360 \qquad + T_2' \qquad)/4 \end{aligned} \tag{5.2.18}$$

or

$$\begin{aligned} T_1' &= (T_B + T_H + T_A \qquad + T_2 \qquad)/4 \\ T_2' &= (T_C + T_G \qquad + T_1' \qquad + T_3)/4 \\ T_3' &= (T_D + T_F + T_E \qquad + T_2' \qquad)/4 \end{aligned} \tag{5.2.19}$$

These differ from eqns. (5.2.11) and (5.2.13) in only one respect: some primed variables occur on the right-hand sides of the update formulae, thus indicating that, assuming the update sequence is T_1, T_2 and T_3, new values are used as soon as they are available. The three operations perform one iteration over the field of the solution domain; and when each iteration is complete, the new variables are treated as improved but non-primed approximations to the solution, and another iteration is performed. There is now no need to store the primed values, as new values computed may be placed immediately in their non-primed locations, and computer programs are generally simpler to construct than their Jacobi counterparts.

First, however, consider how the process works by simple tabulation. The cyclic application of eqn. (5.2.18), starting as before from the trivial guess

TABLE 5.2

GAUSS–SEIDEL ITERATION FOR THE PROBLEM OF FIG. 5.2

Iteration number	T_1	T_2	T_3
0	0	0	0
1	100	125	371·3
2	131·3	225·6	396·4
3	156·4	238·2	399·6
4	159·6	239·8	399·9

$T_1 = T_2 = T_3 = 0$, is illustrated in Table 5.2. It can be seen that after only four iterations, the approximate solution is

$$T_1 = 159{\cdot}6 \quad (160)$$

$$T_2 = 239{\cdot}8 \quad (240)$$

$$T_3 = 399{\cdot}9 \quad (400) \tag{5.2.20}$$

with the exact solution shown in parentheses for comparison.

PROGRAM 2.4 (see p. 254)

A simple computer program to accomplish this Gauss–Seidel iteration task is listed as Program 2.4, and it is strongly analogous to Program 2.1: it is introductory, nothing is complicated and it is specific to the task in hand. As anticipated, it is even simpler than Program 2.1, since primed locations are not used and updating immediately replaces the value in the non-primed location.

PROGRAM 2.5 (see p. 254)

In formulating an improved computer program, a 2-D array is introduced for the T values at the 15 locations as was done in Program 2.3. The idea is also introduced now of coupling (or connecting) coefficients a_N, a_S, a_E and a_W or, specifically, AN(I, J), AS(I, J), AE(, J) AW(I, J). These link the effect of the north, south, east and west neighbours on the central node

$P = (\text{I}, \text{J})$. If there is a source term for internal heat generation it is located in s or S(I, J). The general 2-D Poisson equation

$$k\left(\frac{\partial^2 T}{\partial x^2} + \frac{\partial^2 T}{\partial y^2}\right) + H = 0$$

may be taken in finite difference form in a square mesh of side Δ as

$$T(I-1, J) + T(I+1, J) + T(I, J-1) + T(I, J+1)$$
$$- 4T(I, J) + \frac{H}{k}\Delta^2 = 0 \qquad (5.2.21)$$

or by the energy balance method

$$k\frac{A}{\Delta}[T(I-1, J) - T(I, J)] + \text{three other similar terms} + H\Delta^3 = 0 \qquad (5.2.22)$$

Here the symbols all have the usual meanings as introduced in Chapter 3, Section 3.3. The coupling coefficients are merely the values of the a's such that either of these equations (they are both equivalent) is of the form

$$a_N(T_N - T_P) + a_S(T_S - T_P) + a_E(T_E - T_P) + a_W(T_W - T_P) + s_P = 0 \qquad (5.2.23)$$

and s_P is the constant term, arising from the source. Normally, at interior points of a square mesh, the a's may all be taken equal to unity, but at irregular boundaries or non-square mesh interior points this is not the case. Usually the FDE may be taken as, in rearranged form for updating T_P,

$$T_P = \frac{(a_N T_N + a_S T_S + a_E T_E + a_W T_W)}{a_P} + \frac{s_P}{a_P}$$

where

$$a_P = a_N + a_S + a_E + a_W$$

This can also be written in the index notation form as in the program, Program 2.5, which arranges locations for these quantities. However, because of the simplicity of this particular problem being solved, they are not used as effectively as they could be.

5.2.4. Relaxation

The relaxation procedure is similar to the Gauss–Seidel iteration procedure, and it is occasionally confusing since it is current practice to refer to iterative processes as relaxation ones: in all cases one is relaxing estimated values (for which the FDEs do not hold at the nodes) according to some scheme which eventually makes the FDEs hold to a certain degree of accuracy. The similarity with the Gauss–Seidel procedure is that updated values are used as soon as they are calculated; the difference now is that updating is not done sequentially. After each point update, residuals at all points are calculated (only those at the point in question and its immediate neighbours will have changed due to a single point update, provided that only the five-point formula is being used). The next point to update is decided by observing the absolute values of the residuals at all the points.

TABLE 5.3
RELAXATION FOR THE PROBLEM OF FIG. 5.2

Operation	T_1	R_1	T_2	R_2	T_3	R_3
Initial estimates	0	400	0	400	0	1 360
+340 at 3				740	340	0
+185 at 2		585	185	0		185
+146 at 1	146	1		146		
+46 at 3				192	386	1
+48 at 2		49	233	0		49
+12 at 1	158	1		12		
+12 at 3				24	398	1
+6 at 2		7	239	0		7
+1·75 at 1	159·75	0		1·75		
+1·75 at 3		0·9	239·9	0		0·9
Final values	159·75	0·9	239·9	0	399·75	0·9

The point with highest absolute residual is updated. The procedure continues with single point updates, the order being as determined by the magnitudes of the residuals calculated *en route*, and the procedure terminates when all the residuals reach an acceptably low level (recall that they measure the departure from exactness of solution at the point in question). The method is illustrated in Table 5.3 for our particular simple problem, and it may be seen that after only 11 single point relaxations (equivalent in computational effort to less than four iterations of the Gauss–Seidel or Jacobi procedure) all the residuals are less than unity and the process is terminated with approximate solution:

$$T_1 = 159{\cdot}75 \quad (160)$$

$$T_2 = 239{\cdot}9 \quad (240)$$

$$T_3 = 399{\cdot}75 \quad (400)$$

where again the exact solution is shown bracketed for comparison. In this case the relaxation procedure is about as rapid as the Gauss–Seidel procedure in converging to the exact solution of the FDEs. However, in general, the procedure is more rapid.

This technique is not recommended for computerisation because use must be made of the slow maximum residual search capability of the machine which wastes more time than the increased speed of convergence saves. Note that after each update all the values and residuals are recalculated so as to ensure that the maximum residual is found on the next occasion. In a more general case with many node points this is another waste, for only the four neighbours of the updated point will have their values and residuals altered because of the action taken. A program could be modified to alleviate this, though. Since this technique is not recommended for computers, the details are not given. Note, however, that in simple hand calculations a simple eyeball scan is all that is required to determine which is the next point to update, and the efficient procedure can be used to advantage. However, current utilisation of computers has led to a lack of interest in this technique, and more interest in efficient sequential iterative techniques.

5.2.5. Conducting-Rod Analogy for Calculation of Heat Flows (see Dusinberre, 1961)

For a 2-D square network of side Δ the actual heat flow by conduction between adjacent mesh points can be envisaged with the help of Fig. 5.5. The point P suffers heat flows from its N, S, E and W neighbours, and if all five temperatures at these points are in equilibrium, that is, are solutions of the

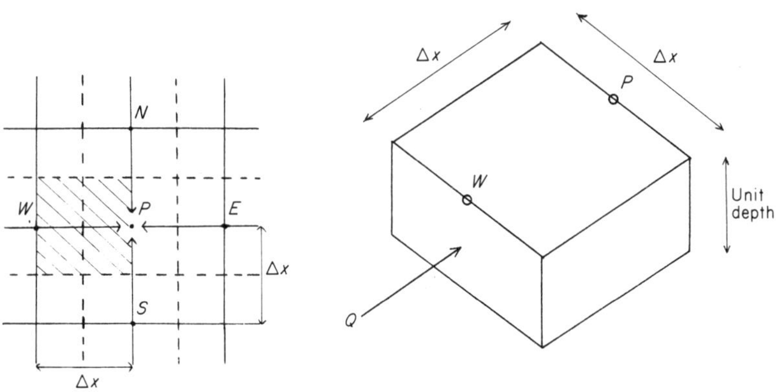

FIG. 5.5. Conducting-rod analogy for calculation of heat flows in a 2-D body.

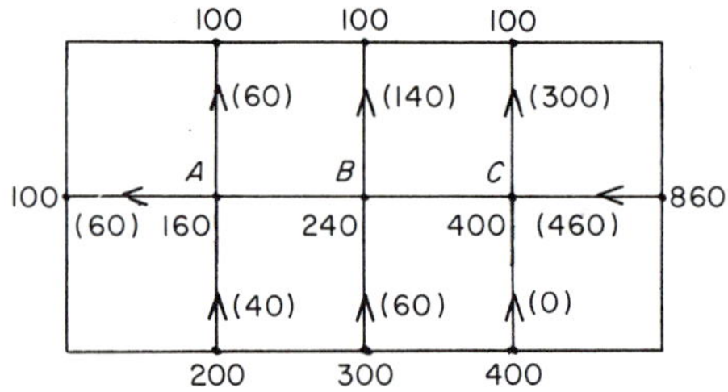

FIG. 5.6. Heat flows for the solution of the simple 2-D problem of Fig. 5.2.

governing FDEs, there is no net heat flow into the cell control volume for P. Taking unit depth of the 2-D object, and using the shaded area in the figure, the heat flow from a typical point W to P is

$$Q_{PW} = k \cdot \frac{(\Delta \cdot 1)}{\Delta}(T_W - T_P)$$
$$= k(T_W - T_P)$$

where k is the conductivity. Similar expressions hold for the other heat flows, as if each pair of adjacent points were connected by conducting rods.

It follows that, by summing the heat flows from the boundaries to the immediate interior points, the total heat flow into or out of the object can be calculated. For the two-dimensional block shown in Fig. 5.6, the rounded-off internal temperatures at the nodes A, B and C are found to be 160°, 240° and 400 °C. Hence

$$\text{Heat inflow} = k(460 + 0 + 60 + 40) = (560k) \text{ W/m length}$$

$$\text{Heat outflow} = k(60 + 60 + 140 + 300) = (560k) \text{ W/m length}$$

The exact balance shows that the relaxation process used to solve the FDEs is complete. Normally there would not be such a precise balance and the magnitude of the difference would be a pointer as to whether further computation was necessary. In this latter case an averaged value of inflows and outflows could be taken.

For the block shown, since the thermal conductivity k is 50 W/mK, the heat transfer per unit length of the block is $560 \times 50 = 2{\cdot}8 \times 10^4$ W/m.

5.3. RELAXATION PROCEDURES

5.3.1. The General Idea

The idea of iterative and relaxation methods for the solution of a simple problem has been outlined in Section 5.2. It is convenient now to put the relaxation ideas in a general context—the relaxation process is a numerical method for solving a set of algebraic equations like the FDEs of heat conduction (see eqn. (3.4.2) of Chapter 3):

$$\sum_j K_{ij}(T_j - T_i) + H_i \cdot V_i = 0 \tag{5.3.1}$$

where the summation is over all j nodes adjacent to the ith node, and there is one such equation for each node i (compare with eqn. (5.2.2)). The summation can be considered to extend over all j nodes in the system of nodes at which the temperature is unknown, in which case all the K_{ij} for which i and j are not adjacent are taken as zero.

In order to solve such a set of equations, like (5.3.1) one such equation for each node of the system, one first assumes a value, a guessed approximate solution of eqn. (5.3.1) based on physical reasoning (it can be as vague as $T_i = 0$). Substituting these values into the left-hand side of eqn. (5.3.1), one finds that the right-hand side is non-zero but has a value R_i which is called the residual at node i:

$$R_i = \sum_j K_{ij}(T_j - T_i) + H_i \cdot V_i \tag{5.3.2}$$

(compare with eqn. (5.2.6)). Any change ΔT_i in T_i for any particular node i will change the R_i's to

$$R_i = R_i - \left(\sum_j K_{ij}\right)\Delta T_i \tag{5.3.3}$$

and any of the R_j's to

$$R_j = R_j + K_{ij}\Delta T_i \tag{5.3.4}$$

where the j's are adjacent to i, or any of the j's if the K_{ij} of such non-adjacent nodes is taken as zero. Thus a change ΔT_i in the temperature T_i at node i causes a redistribution of the residual R_k's which follows eqns. (5.3.3) and (5.3.4). It is the objective of the relaxation procedure to make successive changes ΔT_i in the T_i's so as to reduce all the R_k's to as near zero as possible, this condition representing complete simultaneous solution of eqn. (5.3.1). In the hand relaxation technique, this is accomplished by the following rules:

(i) Compute and record all residuals which result from the assumed values of the temperatures.

(ii) Choose the node i at which the $R_i/\sum_j K_{ij}$ is the largest, and relax it (*i.e.* reduce R_i to zero by changing T_i by an amount equal to $R_i/\sum_j K_{ij}$).

(iii) Compute the new residuals R_j (due to the change of T_i) of the adjacent nodes.

(iv) Again select the largest $R_i/\sum_j K_{ij}$ and repeat steps (ii) and (iii), continuing in this fashion until all residuals are reduced as nearly to zero as accuracy requires.

A systematic recording of the temperatures and residuals is suggested, as shown in the previous (Section 5.2) and following tables, which possess logic and clarity, providing each node i with two columns, one for temperature T_i and one for residual R_i. In the first row are the assumed (guessed) T_i and the corresponding calculated R_i. Any subsequent new row consists of only the changed (new) temperature T_i and the changed (new) values of R_i and R_j for adjacent nodes j.

5.3.2. Over-relaxation to Improve Convergence

If at some stage of the relaxation process there is an uneven distribution of residuals (that is, some positive, some negative) and the basic relaxation technique is followed, the zeroing of the largest absolute residual will increase the absolute value of any adjacent residual of opposite sign. Then the liquidation of these residuals will increase the value of the residual previously zeroed on the previous step. Thus the value of a typical nodal residual oscillates and convergence is slow. This can be avoided if the temperatures at nodes are over-relaxed, so as not merely to zero the appropriate residuals, but to make them take the opposite sign, in expectation that updating nearby nodes on subsequent occasions will have a favourable effect on the over-relaxed residual. One might over-relax perhaps 50 %, but as this topic will be returned to later in Chapter 7, Section 7.3, nothing further is detailed here.

5.3.3. Block Relaxation

Block relaxation is very effective in speeding convergence. The basic relaxation is modified so that instead of reducing only one residual to zero at a particular step, the sum of residuals comprising a particular block is reduced to zero at one go, thus obtaining an even distribution of residuals. In a 2-D square mesh, a single point relaxation of $+1$ in the node temperature changes the node residual by -4 along with a change of $+1$ to the residuals at each of the four neighbouring nodes. In block relaxation, a unit change in the temperatures at each of the nodes in the block is

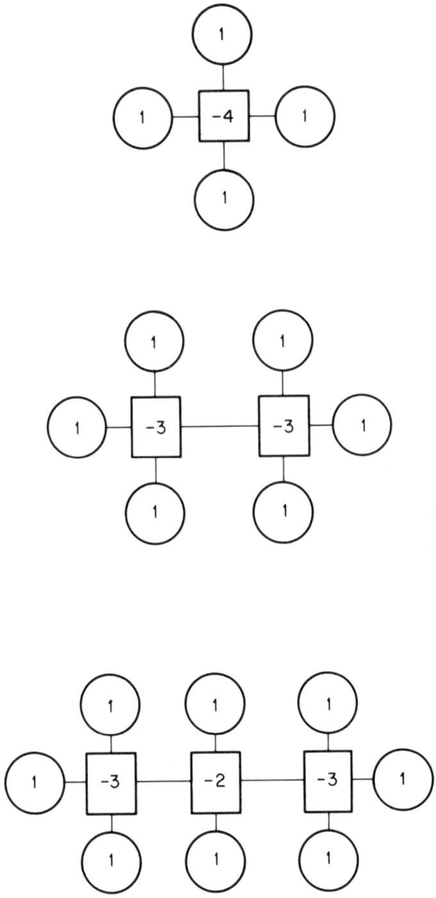

FIG. 5.7. Typical blocks in block relaxation: changes to residuals due to unit change in temperature at each of the nodes within the block.

accompanied by a change in the residuals at nodes within the block by amount -1, -2 or -3 according to the situation along with unit change in the residuals at the surrounding nodes. Typical situations are shown schematically in Fig. 5.7.

The block operator may be deduced by considering logically the effect of two or more simultaneous applications of the single node relaxation operator acting for unit change in temperature at two or more adjacent nodes. Alternatively, note that the sum of the negative numbers associated with nodes within the block is equal to the sum of the positive numbers associated with the nodes surrounding the block, these latter numbers all being + 1. In order to zero the average residual of the nodes inside the block, the sum of these residuals is calculated and divided by 4, 6 or 8 according to which block is to be used. Each block interior node temperature is now changed by this amount. Then each block interior and surrounding residuals are adjusted by the appropriate multiple, given in Fig. 5.7, of this number.

5.3.4. Group Relaxation

Group relaxation is similar to block relaxation, except that changes to the block node temperatures need not be all the same, as was the case informing the block operators shown in Fig. 5.7. One might, for example, choose to adjust the temperatures in a block in proportion to the respective residuals at the nodes in question. In this case one could formulate the appropriate block operator accordingly. For example, if two nodes are adjacent in the E–W direction, and one wishes to give changes of +1 and +2

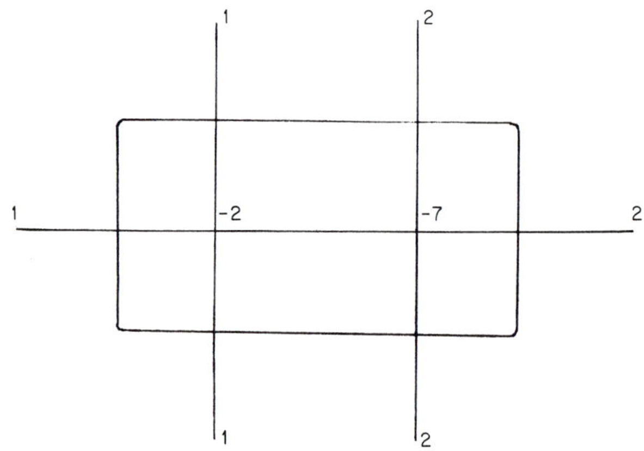

FIG. 5.8. Group relaxation of two adjacent nodes: changes to residuals due to +1 and +2 changes in temperature at the left- and right-hand nodes respectively.

respectively to the temperatures of the left- and right-hand nodes (these being the same proportion as their residuals), then the operator shown in Fig. 5.8 would be appropriate. Success in this method depends very much on the expertise of the operator in deciding what block size, and what block operator, is appropriate to the situation in hand.

5.3.5. Example

A 2-D cartesian coordinate rectangular problem, using a square mesh, is used to illustrate the relaxation solution procedure. The most frequent boundary conditions encountered in practice are those of:

(i) prescribed surface temperature (Dirichlet conditions):
(ii) Newtonian cooling (Neumann conditions);
(iii) prescribed heat flux (for example, Q'' given or adiabatic or lagged boundary conditions).

Figure 5.9 shows a steel slab having prescribed surface temperatures at all the boundaries; this leads to a simpler mathematical problem than those of

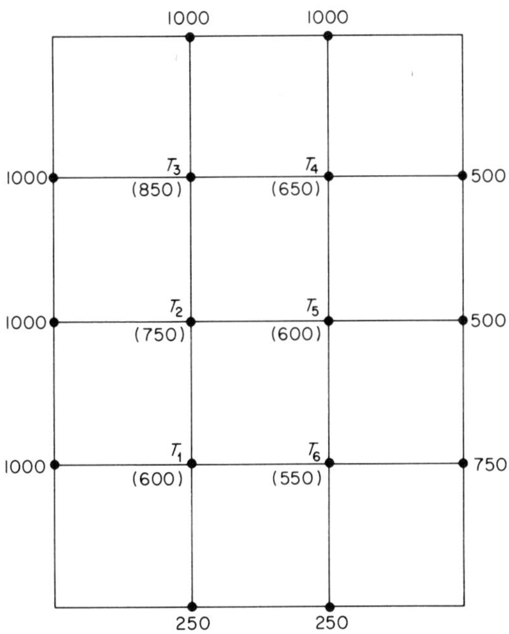

FIG. 5.9. 2-D steady-state steel slab problem with prescribed surface temperatures (Dirichlet conditions).

categories (ii) and (iii). The illustration is modified in the next section, Section 5.4, to show what changes are required to simulate a problem having these extra complexities.

The steel slab has two faces perfectly insulated and the temperature variation along the other faces is as shown in Fig. 5.9. This is therefore a 2-D problem and we shall evolve the steady-state temperature distribution.

TABLE 5.4

TEMPERATURE DISTRIBUTION IN THE 2-D SLAB PROBLEM OF FIG. 5.9 BY RELAXATION

Line	T_1	R_1	T_2	R_2	T_3	R_3	T_4	R_4	T_5	R_5	T_6	R_6
1	600	150	750	50	850	0	650	350	600	50	550	0
2		150		50		88	738	−2		138		0
3	638	−2		88		88		−2		138		38
4		−2		123		88		33	635	−2		73
5		29	781	−1		119		33		29		73
6		29		29	880	−1		63		29		73
7		47		29		−1		63		47	568	1
8		47		29		15	754	−1		63		1
9		47		45		15		15	651	−1		17
10	650	−1		57		15		15		−1		29
11		13	795	1		29		15		13		29
12		13		8	887	1		22		13		29
13		20		8		1		22		20	575	1
14		20		8		6	759	2		25		1
15		20		14		6		8	657	1		7
16	655	0		19		6		8		1		12
17		5	800	−1		11		8		6		12
18		8		−1		11		8		9	578	0
19		8		2	890	−1		11		9		0
20		8		2		2	762	−1		12		0
21		8		5		2		2	660	0		3
22	657	0		7		2		2		0		5
23		2	802	−1		4		2		2		5
24		3		−1		4		2		3	579	1
25		3		0	891	0		3		3		1
26		3		1		0		4	661	−1		2
27		3		1		1	763	0		0		2
28	658	−1		2		1		0		0		3
29		0		2		1		0		1	580	−1
	658		802		891		763		661		580	

There is no internal heat generation. Employing a square mesh system will generate the following set of equations:

$$2000 + T_4 + T_2 - 4T_3 = 0$$
$$1500 + T_3 + T_5 - 4T_4 = 0$$
$$1000 + T_3 + T_5 + T_1 - 4T_2 = 0$$
$$500 + T_2 + T_4 + T_6 - 4T_5 = 0$$
$$1250 + T_2 + T_6 - 4T_1 = 0$$
$$1000 + T_1 + T_5 - 4T_6 = 0$$

This equation set is solved by straightforward relaxation as shown in Table 5.4. There is no need to offer an operations table as long as one keeps an eye on Fig. 5.9, during the course of each temperature update, so as to observe at once which neighbouring residuals are affected.

The initial guesses are seen in Fig. 5.9 (in parentheses at the relevant node). Better initial guesses could have been obtained by considering one node at a time. One assumes first that there is only one interior node and proceeds to calculate its temperature by the heat balance approach. Extra interior points are then inserted one at a time, solving a heat balance for their temperatures as they are inserted, on the assumption that previously calculated values are fixed. In this way the set of initial estimates can be very good indeed.

The calculations shown in Table 5.4 possess none of the refinements of the relaxation method which could be used: over-relaxation, block or group relaxation. Termination is when each residual is less than 3 in absolute magnitude, for then any further node temperature shift would be less than 0·5 °C on the next occasion. The result is

$$T_1 = 658\,^\circ\mathrm{C} \qquad T_4 = 763\,^\circ\mathrm{C}$$
$$T_2 = 802\,^\circ\mathrm{C} \qquad T_5 = 661\,^\circ\mathrm{C}$$
$$T_3 = 891\,^\circ\mathrm{C} \qquad T_6 = 580\,^\circ\mathrm{C}$$

As was discussed under conducting-rod analogy in Section 5.2, it is now possible to calculate heat flows. There are ten rods connecting the six interior points to the boundary, and these are shown in Fig. 5.10 with the computed temperatures and arrows indicating the direction of heat flow in each of the ten cases.

$$\begin{aligned}\text{Heat inflow} &= [(1000 - 658) + (1000 - 802) + (1000 - 891)\\ &\quad + (1000 - 891) + (1000 - 763) + (750 - 580)]k\\ &= 1165k \quad \text{W/m height}\end{aligned}$$

$$\begin{aligned}\text{Heat outflow} &= [(763 - 500) + (661 - 500) + (580 - 250)\\ &\quad + (658 - 250)]k\\ &= 1162k \quad \text{W/m height}\end{aligned}$$

The flow of heat into and out of the material agrees very well, indicating a near solution of the FDEs.

Table 5.5 illustrates how the same problem can be computed using block relaxation. Obviously the same answer as before has been achieved in half

TABLE 5.5

TEMPERATURE DISTRIBUTION IN THE 2-D SLAB PROBLEM OF FIG. 5.9 BY BLOCK RELAXATION

Line	T_1	R_1	T_2	R_2	T_3	R_3	T_4	R_4	T_5	R_5	T_6	R_6	*Block*
1	600	150	750	50	850	0	650	350	600	50	550	0	Initial guess
2		150		50		88	738	−2		138		0	4
3	636	42	786	−22	886	−20		34		174		36	1, 2 and 3
4		42		22		−20		78	644	−2		80	5
5	656	−19		42		−20		78		18	570	19	1 and 6
6		−19		42		0	758	−2		38		19	4
7		−6	799	2		13		11	657	−2		32	2 and 5
8		−6		6	890	1	762	−1		2		32	3 and 4
9		+2		6		1		−1		10	578	0	6
10		5	802	−2		4		2	660	2		3	2 and 5
11	657	1		−1		5		3		3	579	−1	1 and 6
12		2		0		6	763	0	661	0		0	4 and 5
13		2		1	891	2		1		0		0	3
	657		802		891		763		661		579		Final temperatures

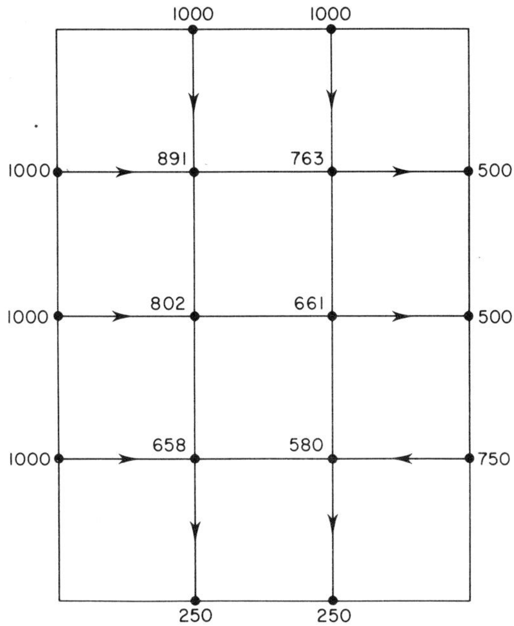

FIG. 5.10. Solution and heat flows for the 2-D slab problem of Fig. 5.9.

the number of lines, but the procedure requires practice to decide when to use block relaxation on a particular line, and also which nodes to block.

5.4. SOME COMPLICATIONS

5.4.1. Convective Boundaries

The simplest way to formulate the FDE for a boundary node B (say) of unknown temperature, where convective boundary effects are to be simulated, is to construct the heat balance for the cell control volume for B. The details have been given in Section 4.3 of Chapter 4 for two cases of special interest:

(i) When B is a general point on a western boundary (see Fig. 4.13 and recall the FDE which was derived for B as eqn. (4.3.11) of Chapter 4):

$$T_B = \frac{\frac{1}{2}(T_N + T_S) + T_E + \mathrm{Bi}\,T_A}{2 + \mathrm{Bi}} \tag{5.4.1}$$

in terms of the Biot number $\mathrm{Bi} = h\,\Delta/k$. Given h, k and the square mesh

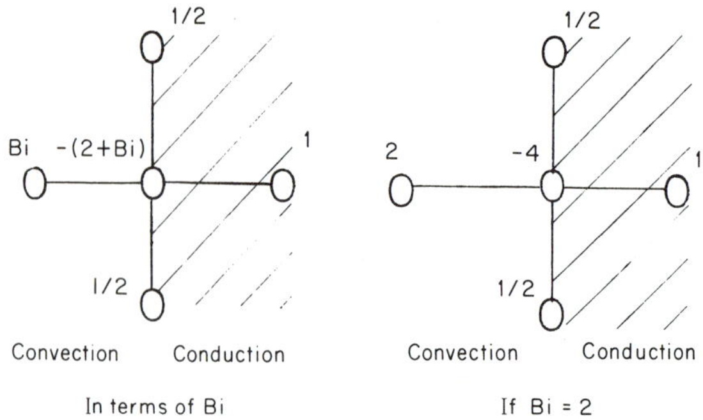

FIG. 5.11. FDE point operators for a western face convective boundary.

spacing Δ, one can substitute for Bi and hence use the simple FDE for the boundary node. If Bi = 2, for example, then the appropriate equation is

$$\tfrac{1}{2}T_N + \tfrac{1}{2}T_S + T_E + 2T_A - 4T_B = 0 \qquad (5.4.2)$$

These FDE point operators for a boundary node on a western boundary can be represented schematically as in Fig. 5.11. Obvious analogies are appropriate for boundary points on north, south or east boundaries.

(ii) When B is a boundary point, not on a side but on a corner. For example, if B occurs at a south-west corner (see Fig. 4.14 and recall the FDE which was derived for B as eqn. (4.3.13) of Chapter 4):

$$T_B = \frac{\tfrac{1}{2}(T_N + T_E) + \mathrm{Bi}\,T_A}{1 + \mathrm{Bi}} \qquad (5.4.3)$$

in terms of the Biot number $\mathrm{Bi} = h\Delta/k$. Given a specific value for Bi, a

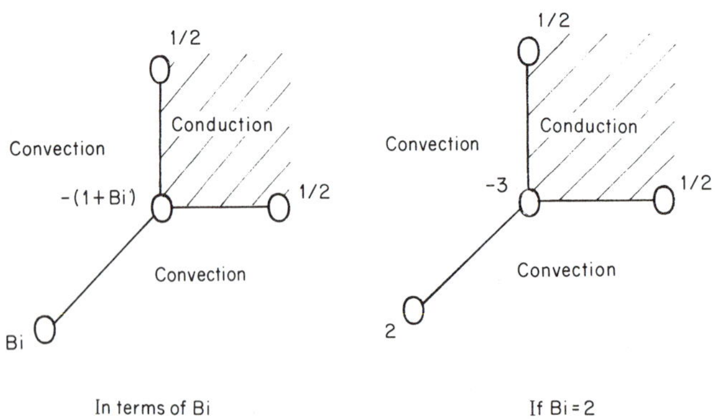

FIG. 5.12. FDE point operators for a south-west corner convective boundary.

simple FDE for such a point follows. If Bi = 2, for example, the appropriate equation is

$$\tfrac{1}{2}T_N + \tfrac{1}{2}T_E + 2T_A - 3T_B = 0 \tag{5.4.4}$$

These FDE point operators for such a south-west boundary node are represented schematically in Fig. 5.12. Again, obvious analogies are appropriate for corners with other configurations.

Using these ideas of eqns. (5.4.1) to (5.4.4) allows the relevant equations at several points in a cross-section of a furnace wall to be deduced. Figure

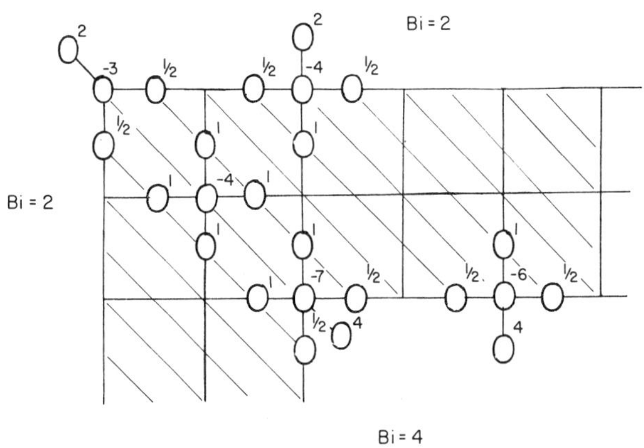

FIG. 5.13. FDEs at several points in a cross-section of a furnace wall with convective boundaries, covered with a square mesh.

5.13 shows the details for such a wall covered with a square mesh and having convective boundaries with inner and outer Biot numbers of 4 and 2 respectively. When the equations for all the nodes are written down, they can be solved by direct, iterative or relaxation methods as already described earlier in this chapter and as they will be described in more detail in Chapters 7 and 8. A case-study problem similar to the furnace wall one just described is discussed in Chapter 8, Section 8.1.

5.4.2. Triangular Mesh

Figure 5.14 shows the cross-section of a furnace. The inside wall is exposed to a gas temperature of T_A through a surface heat transfer coefficient of h. A square mesh of side length Δ is used to cover the wall as shown. Considering the problem as 2-D steady-state, the task is to formulate the FDEs for the nodes labelled 1 to 9. All these except nodes 2

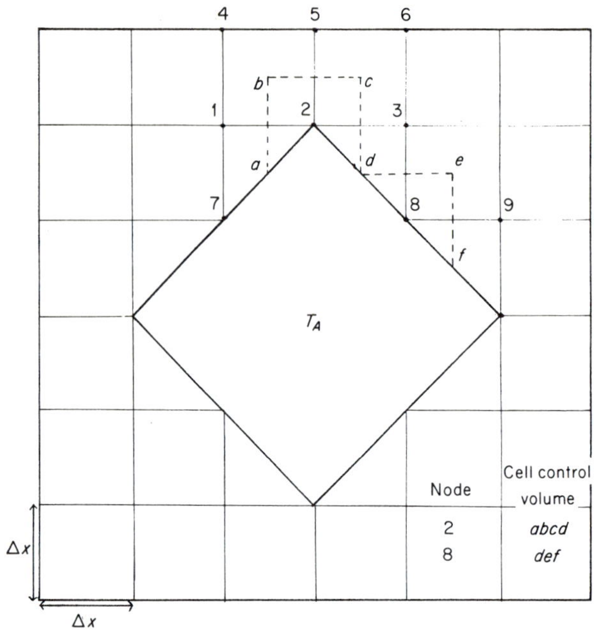

FIG. 5.14. Cross-section of a furnace wall with convective inner boundary, covered with a square mesh.

and 8 can be handled by the conventional methods already described. Here emphasis is placed solely on nodes 2 and 8, where the mesh reveals that triangular segments are to be used adjacent to the inner convective boundary.

Node 2. Considering triangle 238, the geometric factors F_{ij}, which are conductances divided by conductivity K_{ij}/k (see Chapter 4, Section 4.2), are given by

$$F_{82} = \tfrac{1}{2}\cot 90^\circ$$

$$F_{32} = \tfrac{1}{2}\cot 45^\circ$$

that is,

$$F_{82} = 0$$

$$F_{32} = \tfrac{1}{2}$$

Similarly

$$F_{72} = 0$$

$$F_{12} = \tfrac{1}{2}$$

The remainder of the furnace section can be mapped adequately with a square mesh in the usual manner. There will thus be further contributions to the energy balance of node 2 from nodes 1, 3 and 5. From squares 1245 and 2563 there are additional geometric factors:

$$F_{12} = \tfrac{1}{2}$$

$$F_{52} = 1$$

$$F_{32} = \tfrac{1}{2}$$

Thus the final geometric factors affecting node 2 are

$$F_{12} = \tfrac{1}{2} + \tfrac{1}{2} = 1$$

$$F_{52} = 1$$

$$F_{32} = \tfrac{1}{2} + \tfrac{1}{2} = 1$$

The FDE for node 2 can now be written down, assuming constant conductivity k, as

$$kF_{12}(T_1 - T_2) + kF_{52}(T_5 - T_2) + kF_{32}(T_3 - T_2) + h(2)^{1/2}\Delta(T_A - T_2) = 0$$

which reduces to

$$T_1 + T_3 + T_5 + (2)^{1/2}\mathrm{Bi}T_A - [3 + (2)^{1/2}\mathrm{Bi}]T_2 = 0 \qquad (5.4.5)$$

where the Biot number $\mathrm{Bi} = h\Delta/k$.

Node 8. For node 8 the FDE can be derived in a similar fashion, leading to

$$T_3 + T_9 + (2)^{1/2}\mathrm{Bi}T_A - [2 + (2)^{1/2}\mathrm{Bi}]T_8 = 0 \qquad (5.4.6)$$

BIBLIOGRAPHY

AMES, W. F. (1969). *Numerical Methods for Partial Differential Equations*. Nelson, London.

CARNAHAN, B., LUTHER, H. A. & WILKES, J. O. (1969). *Applied Numerical Methods*. Wiley, New York.

CARSLAW, H. S. & JAEGER, J. C. (1959). *Conduction of Heat in Solids*. Clarendon Press, Oxford.

DORN, W. S. & MCCRACKEN, D. D. (1972). *Numerical Methods with Fortran IV: Case Studies*. Wiley, New York.

DUSINBERRE, G. M. (1961). *Heat Transfer Calculations by Finite Differences*. International Textbook Company, Scranton, Pa.

FORSYTHE, G. E. & WASOW, W. R. (1960). *Finite Difference Methods for Partial Differential Equations*. Wiley, New York.

FOX, L. (1962). *Numerical Solution of Ordinary and Partial Differential Equations*. Pergamon Press, Oxford; Addison-Wesley, Reading, Mass.

GARDNER, K. (1945). Efficiency of extended surface. *Trans. ASME*, **67**(8), pp. 621–31.

GERALD, C. F. (1970). *Applied Numerical Analysis*. Addison-Wesley, Reading, Mass.

MCCRACKEN, D. D. (1965). *A Guide to Fortran IV Programming*. Wiley, New York.

MITCHELL, A. R. (1969). *Computational Methods in Partial Differential Equations*. Wiley, New York.

RAZELOS, P. (1973). Methods of obtaining approximate solutions. Section 4, pp. 4–1 to 4–78, in *Handbook of Heat Transfer*, ed. W. M. Rohsenow and J. P. Hartnett. McGraw-Hill, New York.

REID, J. K. (1966). A method for finding the optimum successive over-relaxation parameter. *Comput. J.*, **9**, p. 200.

RICHTMYER, R. D. & MORTON, K. W. (1967). *Difference Methods for Initial Value Problems*, 2nd ed. Interscience, New York.

SMITH, G. D. (1965). *Numerical Solution of Partial Differential Equations*. Oxford University Press, London.

VARGA, R. S. (1962). *Matrix Iterative Analysis*. Prentice-Hall, Englewood Cliffs, NJ.

CHAPTER 6

Application of Finite Difference Equations to Simple Transient Problems

6.1. 1-D TRANSIENT PROBLEMS: EXPLICIT SCHEME

6.1.1. Simple Problem

Consider the 1-D problem shown in Fig. 6.1 of a steel slab, initially at 20 °C throughout, which has both faces suddenly increased to, and subsequently maintained at, 1000 °C. The slab is large in dimensions perpendicular to its thickness, which measures 210 mm across. Given the following properties of steel for conductivity, specific heat and density:

$$k = 41\ \mathrm{W/m\,K}$$
$$C_p = 504\ \mathrm{J/kg\,K}$$
$$\rho = 8000\ \mathrm{kg/m^3}$$

the problem is to determine the temperature distribution throughout the slab at a time 5 min after the initial temperature change took place. The first step is to subdivide the object into a number of segments; six divisions will produce a segment thickness of 35 mm as shown in Fig. 6.1. The next step is to derive the FDEs for all the nodes and then decide how many times to

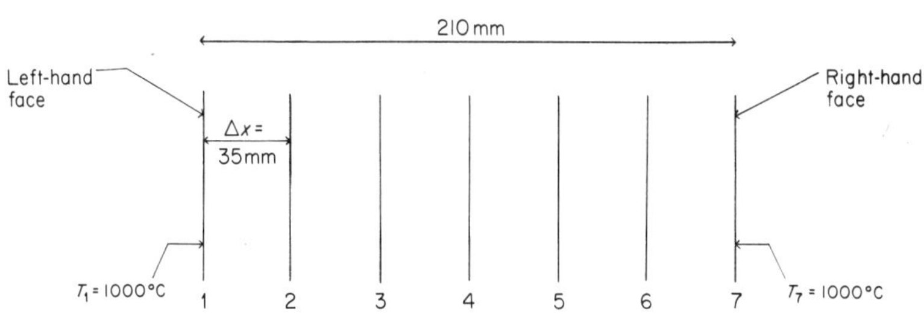

FIG. 6.1. Subdivision of 1-D steel slab into six slices for transient numerical computation.

apply the FDEs across the object so as to obtain the temperature distribution after 5 min.

The FDE for node 2, for example, can be obtained directly from eqn. (3.5.18) of Chapter 3, which when modified for 1-D flow gives

$$\sum_j K_{j2}(T_j - T_2) + HV_2 = \rho C_p V_2 \frac{\Delta T_2}{\Delta t} \tag{6.1.1}$$

where the summation is over neighbouring nodes, K_{j2} is the conductance from node j to node 2 (see eqn. (3.5.15) of Chapter 3), H is the heat generation rate (W/m^3), V_2 is the cell control volume associated with node 2, and $\Delta T_2 = T_2' - T_2$, which is the temperature change at node 2 as time goes from t to $t + \Delta t$.

In this first problem, H is zero, V_2 is $(1.1.\Delta x)$ and the summation is over adjacent nodes, $j = 1$ and 3 only. Hence

$$k\frac{1^2}{\Delta x}(T_1 - T_2) + k\frac{1^2}{\Delta x}(T_3 - T_2) = \rho C_p \Delta x \frac{(T_2' - T_2)}{\Delta t}$$

which leads to

$$T_1 + T_3 - \left(2 - \frac{1}{\mathrm{Fo}}\right)T_2 = \frac{1}{\mathrm{Fo}}T_2' \tag{6.1.2}$$

where the Fourier number

$$\mathrm{Fo} = \alpha\frac{\Delta t}{(\Delta x)^2} = \frac{k}{\rho C_p}\frac{\Delta t}{(\Delta x)^2} \tag{6.1.3}$$

This equation is precisely eqn. (3.5.5) of Chapter 3 which was obtained by direct finite difference approximation of the governing 1-D cartesian transient PDE

$$\frac{\partial^2 T}{\partial x^2} = \frac{1}{\alpha}\frac{\partial T}{\partial t} \tag{6.1.4}$$

viz.

$$\frac{T_1 - 2T_2 + T_3}{(\Delta x)^2} = \frac{1}{\alpha}\left(\frac{T_2' - T_2}{\Delta t}\right)$$

from which eqn. (6.1.2) follows at once.

The edge boundary conditions are of Dirichlet type, that is, known values of temperature. Hence the stability criterion of the explicit representation is the avoidance of negative coefficients in eqn. (6.1.2) (see eqn. (4.5.2) of Chapter 4):

$$\mathrm{Fo} \leq 0{\cdot}5 \tag{6.1.5}$$

for the 1-D case considered here. Taking

$$\mathrm{Fo} = 0{\cdot}5 \tag{6.1.6}$$

the time-step Δt must be taken as

$$\Delta t = 60{\cdot}2\,\mathrm{s}$$

by virtue of eqn. (6.1.3) and then the FDE (eqn. (6.1.2)) simplifies to

$$T_2' = \tfrac{1}{2}(T_1 + T_3)$$

A similar equation holds for each of the other interior nodes at which T is unknown:

$$T_i' = \tfrac{1}{2}(T_{i-1} + T_{i+1}) \quad (2 \leq i \leq 6) \tag{6.1.7}$$

and the application of this to each of the five interior points yields temperatures at the next time-level, 60·2 s later. In this way, with subsequent repetition, eqn. (6.1.7) allows the solution to be marched forward in time as far as required. The calculations for five time-steps are shown in Table 6.1, five time-steps corresponding to 301 s or approximately 5 min. If values are required especially at some particular time, one can either

(i) use linear interpolation between values at two time-levels (see the later illustration); or
(ii) reduce Δt, thus reducing Fo and so maintaining the stability criterion, so that an integer multiple of Δt gives the required time exactly.

TABLE 6.1
STEEL SLAB TEMPERATURE RESPONSE WITH FOURIER NUMBER Fo = 0·5
(Each time-step represents approx. 1 min)

Time-step	*t (min), approx.*	*Left-hand face* T_1	T_2	T_3	T_4	T_5	T_6	*Right-hand face* T_7
0	0	1 000	20	20	20	20	20	1 000
1	1	1 000	510	20	20	20	510	1 000
2	2	1 000	510	265	20	265	510	1 000
3	3	1 000	632·5	265	265	265	632·5	1 000
4	4	1 000	632·5	448·8	265	448·8	632·5	1 000
5	5	1 000	724·4	448·8	448·8	448·8	724·4	1 000

PROGRAM 1.1 (see p. 256)

The values calculated and shown in Table 6.1 could equally well be obtained via a computer. A simple computer program which will calculate the time–temperature history for the problem of Fig. 6.1 is given as Program 1.1. The comment cards provide adequate description of the arrangement of the computation.

In the table, the initial face temperatures are taken as 1000 °C, which is their temperature when $t > 0$, rather than their initial value of 20 °C at $T = 0$. This is one way of overcoming the temperature discontinuity at $t = 0$, justified by the assumption that during the first time-step of 60·2 s this is the temperature of the surface which is affecting the slab temperature response. A better recommendation is to give such points their mean value at their discontinuity, taking, in our case,

$$T_1 = T_7 = 510\,^\circ\mathrm{C}$$

This is in direct conflict with the explicit representation of taking initial values over a subsequent time-step, but in fact leads to more physically accurate solutions. Calculations with this initial face temperature are shown

TABLE 6.2

STEEL SLAB TEMPERATURE RESPONSE WITH FOURIER NUMBER Fo = 0·5 AND INITIAL FACE DISCONTINUITIES GIVEN MEAN VALUES

(Each time-step represents approx. 1 min)

Time-step	*Left- or right-hand face* T_1 $(= T_7)$	T_2 $(= T_6)$	T_3 $(= T_5)$	T_4	T_5 $(= T_3)$
0	510	20	20	20	20
1	1 000	265	20	20	20
2	1 000	510	142·5	20	142·5
3	1 000	571·3	265	142·5	265
4	1 000	632·5	356·9	265	356·9
5	1 000	678·5	448·8	356·9	448·8

in Table 6.2, where advantage is also taken of the symmetry condition, that halfway across the slab

$$\frac{\partial T}{\partial x} = 0$$

Hence

$$T_5 = T_3; \quad T_6 = T_2; \quad T_7 = T_1$$

Accordingly the FDE for node 4, the centre node, is eqn. (6.1.7) with $T_5 = T_3$, that is, double weighting is attached to node 3 in the FDE for node 4. Alternatively one can just copy the T_3 figure into T_5 locations and apply eqn. (6.1.7) at node 4 in the usual way, this being used in the version shown in the table.

PROGRAM 1.2 (see p. 257)

A simple computer program to do these calculations and obtain the results shown in Table 6.2 is given as Program 1.2. Again comment cards provide adequate information about the arrangement of the program. Notice also how the symmetry condition $\partial T/\partial x = 0$ at node 4 is used to reduce the computational effort.

6.1.2. Illustration in Cylindrical Polar Coordinates

Consider a 1-D cylindrical polar coordinate problem (shown in Fig. 6.2) in which z and θ variations are insignificant. Assume that the outer

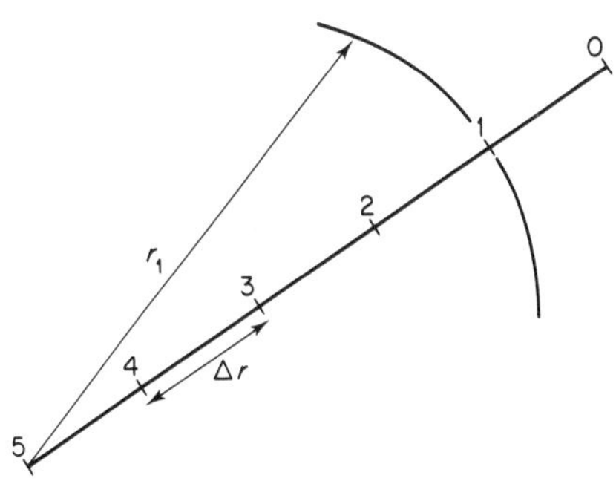

FIG. 6.2. 1-D cylindrical polar coordinate problem.

boundary is lagged and that internal heat generation H is operative. The relevant PDE from Table 2.5 of Chapter 2 with the addition of an internal heat generation term is

$$k\left(\frac{\partial^2 T}{\partial r^2} + \frac{1}{r}\frac{\partial T}{\partial r}\right) + H = \rho C_p \frac{\partial T}{\partial t} \tag{6.1.8}$$

The discussion in Section 4.1 of Chapter 4 focused on appropriate FDEs for equations similar to this one. In particular, the appropriate form of eqn. (4.1.6) of Chapter 4 with the z and θ variations suppressed and the addition of a heat generation term HV_0, where V_0 is the cell control volume for the node 0 given by eqn. (4.1.4) of Chapter 4, is all right for the left-hand side of eqn. (6.1.8). The right-hand side requires a term like eqn. (4.1.5) of Chapter 4. Some of these details are included in the appropriate part of Table 4.2 of Chapter 4. The figure includes a nodal subdivision taking five sections of equal Δr spacing.

The appropriate FDE may also be obtained by direct PDE replacement of eqn. (6.1.8); the result is

$$k\left[\frac{T_0 - 2T_1 + T_2}{(\Delta r)^2} + \frac{1}{r_1}\frac{(T_0 - T_2)}{2\Delta r}\right] + H = \rho C_\mathrm{p}\frac{(T_1' - T_1)}{\Delta t}$$

for node 1, and similar expressions for each of the other nodes, except the centre node. In terms of the Fourier number Fo where

$$\mathrm{Fo} = \alpha\frac{\Delta t}{(\Delta r)^2} = \frac{k}{\rho C_\mathrm{p}}\frac{\Delta t}{(\Delta r)^2}$$

this equation, written for a general interior node i, becomes

$$T_i' = \frac{\mathrm{Fo}}{2}\left[\left(2 + \frac{\Delta r}{r_i}\right)T_{i-1} + \left(\frac{2}{\mathrm{Fo}} - 4\right)T_i + \left(2 - \frac{\Delta r}{r_i}\right)T_{i+1} + \frac{2H}{k}(\Delta r)^2\right] \tag{6.1.9}$$

For the centre node 5 it has been shown (Section 4.1 of Chapter 4) that

$$\left.\frac{1}{r}\frac{\partial T}{\partial r}\right|_{r\to 0}$$

can be replaced by

$$\left.\frac{\partial^2 T}{\partial r^2}\right|_{r\to 0}$$

so that an appropriate FDE for the centre node 5 is

$$2k\left\{\frac{2T_4 - 2T_5}{(\Delta r)^2}\right\} + H = \rho C_\mathrm{p}\frac{(T_5' - T_5)}{\Delta t}$$

which reduces to

$$T_5' = \mathrm{Fo}\left[4T_4 + \left(\frac{1}{\mathrm{Fo}} - 4\right)T_5 + \frac{H}{k}(\Delta r)^2\right] \tag{6.1.10}$$

where the symmetry condition

$$\left.\frac{\partial T}{\partial r}\right|_{r=0} = 0$$

has been used. A simpler alternative is to compute T_4' in the usual fashion, using eqn. (6.1.9), and then simply set

$$T_5' = T_4' \tag{6.1.11}$$

but since this assumes more than the symmetry condition, viz. it assumes also that

$$\left.\frac{\partial T}{\partial r}\right|_{r = \Delta r/2} = 0$$

it is less accurate than the use of eqn. (6.1.10). Accordingly it is recommended to use eqn. (6.1.10) for the centre point, node 5, and equations analogous to eqn. (6.1.9) at other interior points, labelled 1 to 4.

In order to handle the zero normal gradient condition at the outer boundary, node 0, it is simplest in PDE replacement techniques to introduce a fictitious node F distant Δr further out than node 0 at radius $(r_0 + \Delta r)$. The usual equation, eqn. (6.1.9), is applied to node 0 together with the zero normal gradient condition

$$\left.\frac{\partial T}{\partial r}\right|_{r = r_0} = 0$$

that is,

$$\frac{T_F - T_1}{2\Delta r} = 0$$

which reduces simply to

$$T_F = T_1$$

Thus the equation for the outer node 0 is

$$T_0' = \mathrm{Fo}\left[2T_1 + \left(\frac{1}{\mathrm{Fo}} - 2\right)T_0 + \frac{H}{k}(\Delta r)^2\right] \tag{6.1.12}$$

Solution of the transient problem is thus obtained by marching forward in time in steps of Δt, advancing the temperature at the centre node 5 using eqn. (6.1.10), at interior nodes 4, 3, 2, and 1 using eqn. (6.1.9) and at the outer boundary node 0 using eqn. (6.1.12). Inspection of all the right-hand sides of these equations reveals that the stability criterion of avoiding negative coefficients is maintained provided $\mathrm{Fo} \leq 0{\cdot}25$, this restriction downward from 0·5 arising because of the treatment of the centre node.

In case the reader is deliberating on whether a logarithmic subdivision in the r-direction would simplify the nodal equations, it would in the region where it was applied. These modifications have not been applied here, though, since in any case the region cannot be extended to include the centre node at $r = 0$, as has been remarked already in Section 4.1 of Chapter 4, and thus there would be a matching problem at the first interior node adjacent to the centre node. It would thus be inconvenient to keep changing from ξ to r as independent variable (see eqn. (4.1.12) of Chapter 4).

6.2. ALTERING MESH SIZE AND INSTABILITY

According to the discussion in Section 4.5 of Chapter 4, provided the FDE representation is convergent, decreasing Δx and/or Δt will improve the accuracy. One must be careful of course to ensure that the stability criterion, eqn. (6.1.5) for the simple problem of Section 6.1, is still adhered to when Δx is refined without a corresponding refinement of Δt. In fact if Fo = 0·5 and Δx is halved, Δt needs to be reduced to a quarter of its value in order to keep Fo = 0·5 and maintain stability. If Δt is not reduced by at least this much, Fo > 0·5 and the process would be unstable, as illustrated later in Table 6.5. Hence it is seen that halving Δx must be accompanied by at least quartering Δt, leading to a multiple of 8 in the work-load required to reach a particular time.

The effect of increasing the number of slices from 6 to 8 for the simple problem of Section 6.1 is illustrated in Table 6.3. The Fourier number Fo is maintained at 0·5, for which Δt must now be taken as 33·95 s. To obtain the solution after 5 min, interpolation is required between time-steps 8 and 9; in fact, since

$$\frac{300}{33{\cdot}95} = 8{\cdot}86$$

one can use linear interpolation in the form

$$T_i|_{t=300} = 0{\cdot}14T_i|_{\text{time-level } 8} + 0{\cdot}86T_i|_{\text{time-level } 9} \tag{6.2.1}$$

Since this slight refinement of Δx and Δt should improve the accuracy, one would expect the centre-line value (T_5 of Table 6.3) after 5 min to be more accurate than its corresponding value with the coarser grid (T_4 of Table 6.2). One would therefore expect 364·7 °C to be more realistic than 356·9 °C.

Another way to improve the accuracy, rather than refining Δx and Δt simultaneously, is to refine Δt in isolation, thus reducing the Fourier number Fo and therefore maintaining stability. If, for example, the number of time-steps is increased from 5 to 8, while maintaining the number of slabs at 6, then Δt reduces to $60{\cdot}2 \times 0{\cdot}625 = 37{\cdot}5$ s and Fo reduces to $0{\cdot}5 \times 0{\cdot}625 = 0{\cdot}312$. Thus the FDE to be solved is slightly more complicated

TABLE 6.3

STEEL SLAB TEMPERATURE RESPONSE WITH FOURIER NUMBER Fo = 0·5 WITH REFINED Δx SPACING

(Each time-step represents 33·95 s)

Time-step	T_1	T_2	T_3	T_4	T_5	$T_6\ (= T_4)$
0	510	20	20	20	20	20
1	1 000	265	20	20	20	20
2	1 000	510	142·5	20	20	20
3	1 000	571·3	265	81·25	20	81·25
4	1 000	632·5	326·3	142·5	81·25	142·5
5	1 000	663·2	387·5	203·8	142·5	203·8
6	1 000	693·8	433·5	265	203·8	265
7	1 000	716·8	479·4	318·6	265	318·6
8	1 000	739·7	517·7	372·2	318·6	372·2
9	1 000	758·9	556	418·1	372·2	418·1
Interpolating for the 8·86 time-step yields						
8·86	1 000	756·2	550·6	411·7	364·7	411·7

and is obtained by substituting this value of Fo in eqn. (6.1.2); the result for each general interior node is

$$T_i' = 0{\cdot}312(T_{i-1} + T_{i+1}) + 0{\cdot}376T_i \quad (2 \leq i \leq 6) \qquad (6.2.2)$$

A check on the arithmetic is provided by observing that the coefficients on the right of eqn. (6.2.2) always add to unity, when the coefficient of T_i' on the left is unity. That this is a requirement is easily deduced from eqn. (6.1.2). Application of eqn. (6.2.2) generates results, the final line at $t = 5$ min being as shown in Table 6.4.

TABLE 6.4

STEEL SLAB TEMPERATURE RESPONSE WITH FOURIER NUMBER Fo = 0·312

(Each time-step represents 37·5 s)

Time-step	T_1	T_2	T_3	T_4	$T_5\ (= T_3)$
0	510	20	20	20	20
1	1 000	172·9	20	20	20
2	1 000	383·2	67·7	20	67·7
3	1 000	477·2	151·3	49·8	151·3
4	1 000	538·6	221·3	113·1	221·3
5	1 000	583·6	286·6	180·6	286·6
6	1 000	620·8	346·2	246·7	346·2
7	1 000	653·4	400·8	308·8	400·8
8	1 000	682·8	450·9	366·2	450·9

The computations of Tables 6.3 and 6.4 can easily be accomplished by the second computer program, Program 1.2, via obvious alteration to the DIMENSION statement, DO loop ranges for the number of locations across the slab and for the number of time-steps required. For the values of the latter table, the only other serious modification is the statement within the DO 101 loop; it should be replaced by

```
      DO 101 I = 2,4
      TP(I) = 0·312*(T(I - 1) + T(I + 1)) + 0·376*T(I)
  101 CONTINUE
```

Of course, when refining the Δx and Δt grids for improved accuracy, one must be careful to maintain $\mathrm{Fo} \leq 0{\cdot}5$ to ensure stability. Shown in Table 6.5 is the effect of not doing this, and haphazardly taking $\Delta t = 120$ s with $\Delta x = 3{\cdot}5$ cm (6 slices across the slab) and so using $\mathrm{Fo} = 1$. In this case the FDE is

$$T_i' = T_{i-1} + T_{i+1} - T_i \quad (2 \leq i \leq 6) \tag{6.2.3}$$

and computations with this reveal quite unacceptable features, associated with instability of the FDE formulation—oscillations of increasing magnitude as the calculation proceeds. Linear interpolation between time-steps 2 and 3 to obtain the solution at $t = 5$ min ($= 2{\cdot}5$ time-steps) would clearly not be satisfactory.

TABLE 6.5

STEEL SLAB TEMPERATURE RESPONSE WITH FOURIER NUMBER Fo = 1, SHOWING AN UNSTABLE CALCULATION PROCEDURE

(Each time-step represents 2 min)

Time-step	T_1	T_2	T_3	T_4	$T_5\ (= T_3)$
0	510	20	20	20	20
1	1 000	510	20	20	20
2	1 000	510	510	20	510
3	1 000	1 000	20	1 000	20
4	1 000	20	1 980	−960	1 980
5	1 000	2 960	−2 920	4 920	−2 920

6.3. 2-D TRANSIENT PROBLEMS: EXPLICIT SCHEME

Two-dimensional transient heat flow problems are governed by the PDE

$$k\left(\frac{\partial^2 T}{\partial x^2} + \frac{\partial^2 T}{\partial y^2}\right) + H = \rho C_p \frac{\partial T}{\partial t} \tag{6.3.1}$$

and how to replace this by a set of FDEs has already been considered in Chapter 3, Section 3.5. For simplicity, consider the problem of Fig. 5.2 of Chapter 5 (which was treated there as a steady-state problem), which is that of a rectangular region with sides in the proportion 1:2. A square mesh system is utilised to subdivide the 2-D spatial region in cartesian coordinates, as shown, resulting in only three interior nodes labelled 1, 2 and 3 at which temperatures are to be computed. It is assumed that at all times the boundary points $A, B, \ldots, H$ have known constant temperatures (Dirichlet boundary conditions) as shown in Fig. 5.3 of Chapter 5, but that at time $t = 0$ all interior points of the region are at a temperature $T = 0$. Assuming no internal heat generation ($H = 0$), direct replacement of eqn. (6.3.1) or energy balance techniques show that an explicit FDE formulation is (see eqn. (3.5.4) of Chapter 3 and Table 4.1 of Chapter 4)

$$\frac{1}{\text{Fo}} T'_P = T_N + T_S + T_E + T_W - \left(4 - \frac{1}{\text{Fo}}\right) T_P \tag{6.3.2}$$

where the Fourier number

$$\text{Fo} = \alpha \frac{\Delta t}{\Delta^2} = \frac{k \Delta t}{\rho C_{\text{p}} \Delta^2}$$

and obvious compass notation has been used. Non-primed quantities here are evaluated at the old time-level t; the primed quantity T'_P is to be the newly calculated temperature at P at the new time-level $t + \Delta t$. By virtue of eqn. (4.5.2) of Chapter 4, given ρ, C_{p}, k and Δ, one must choose a time-step Δt such that the Fourier number $\text{Fo} \leq 0{\cdot}25$, which amounts to ensuring no negative coefficients on the right-hand side of eqn. (6.3.2). Thus in order to advance to a new time-level the three temperatures T'_1, T'_2 and T'_3 must be calculated (each such calculation involving T's at the old time-level), using the boundary conditions where required:

$$\frac{1}{\text{Fo}} T'_1 = T_B + T_H + T_2 + T_A - \left(4 - \frac{1}{\text{Fo}}\right) T_1$$

$$\frac{1}{\text{Fo}} T'_2 = T_C + T_G + T_3 + T_1 - \left(4 - \frac{1}{\text{Fo}}\right) T_2$$

$$\frac{1}{\text{Fo}} T'_3 = T_D + T_F + T_E + T_2 - \left(4 - \frac{1}{\text{Fo}}\right) T_3 \tag{6.3.3}$$

Incorporating the known boundary conditions as given in Fig. 5.3 of Chapter 5 leads to

$$\frac{1}{\text{Fo}} T_1' = 400 + T_2 - \left(4 - \frac{1}{\text{Fo}}\right) T_1$$

$$\frac{1}{\text{Fo}} T_2' = 400 + T_3 + T_1 - \left(4 - \frac{1}{\text{Fo}}\right) T_2$$

$$\frac{1}{\text{Fo}} T_3' = 1360 + T_2 - \left(4 - \frac{1}{\text{Fo}}\right) T_3 \qquad (6.3.4)$$

Taking the rectangle to measure 0·1 m × 0·2 m gives the square mesh spacing $\Delta = 0{\cdot}05$ m; and assuming it is constructed of steel with conductivity, specific heat and density given as in Section 6.1 by

$$k = 41 \text{ W/m K}$$
$$C_p = 504 \text{ J/kg K}$$
$$\rho = 8000 \text{ kg/m}^3$$

then the time-step Δt and Fourier number Fo are related by

$$\text{Fo} = \frac{41}{8000 \times 504} \frac{\Delta t}{(0{\cdot}05)^2} \approx 0{\cdot}004\, \Delta t$$

To simplify the calculations, take Fo equal to its upper limit value of 0·25 in eqn. (6.3.3) or (6.3.4). The corresponding Δt is

$$\Delta t = \frac{0{\cdot}25}{0{\cdot}004} = 62{\cdot}5 \text{ s}$$

and the marching equations become

$$T_1' = (T_B + T_H + T_2 + T_A)/4$$
$$T_2' = (T_C + T_G + T_3 + T_1)/4$$
$$T_3' = (T_D + T_F + T_E + T_2)/4 \qquad (6.3.5)$$

or

$$T_1' = (400 \qquad + T_2 \qquad)/4$$
$$T_2' = (400 + T_1 \qquad + T_3)/4$$
$$T_3' = (1360 \qquad + T_2 \qquad)/4 \qquad (6.3.6)$$

The calculation with eqn. (6.3.6) is illustrated in Table 6.6, for 7 time-steps corresponding to 437·5 s, or 7 min 17·5 s. Inspection of the calculated temperatures in Table 6.6 reveals that they follow precisely the same pattern as the Jacobi iteration calculation for the solution of the corresponding steady-state problem, as seen in Table 5.1 of Chapter 5. There each new line represented another iteration towards the steady-state solution; here each new line is associated with a time-step and a definite time en route to the eventual steady-state condition. The resemblance of Table 6.6 to Table 5.1 of Chapter 5 is not merely fortuitous: there is a strong analogy between the

TABLE 6.6
2-D TRANSIENT CALCULATION USING EXPLICIT FORMULATION AND Fo = 0·25

Time-step	*Time t* (sec)	T_1	T_2	T_3
0	0	0	0	0
1	62·5	100	100	340
2	125	125	210	365
3	187·5	152·5	222·5	392·5
4	250	155·6	236·3	395·6
5	312·5	159·1	237·8	399·1
6	375	159·45	239·6	399·45
7	437·5	159·7	239·7	399·9

two processes. In fact the time-march equations, eqns. (6.3.5) and (6.3.6), are exactly the same as the Jacobi iteration equations, eqns. (5.2.11) and (5.2.13) of Chapter 5. This is a general result: the explicit transient calculation of a 1-D, 2-D or 3-D problem with known (Dirichlet) boundary conditions and using the limiting value of the Fourier number Fo (= 1/2, 1/4 or 1/6, respectively) tracks precisely the same calculation as the corresponding steady-state problem solved via the Jacobi iteration procedure. Thus a suitable computer program for the explicit transient problem is the corresponding steady-state program, Program 1.2 of Chapter 4, appropriately modified slightly so that each new iteration is now associated with a new time-level. The details, being so simple, are not given here.

6.4. CONVECTIVE BOUNDARY CONDITIONS

6.4.1. 1-D Simple Problem

In Section 6.1 the simple transient problem of a 1-D steel slab was considered, where Dirichlet boundary conditions were given, so that the temperatures on both sides of the slab were known for all time. The problem is illustrated in Fig. 6.1, where initially the slab is at 20 °C and suddenly the two faces are increased to, and subsequently maintained at, 1000 °C. Tables 6.1 and 6.2 illustrate computations for that problem, which could also be solved using Programs 1.1 and 1.2.

In heat conduction problems the more usual situation at the end-points (faces of the slab in this case) is not that they are held at known constant temperatures, but that heat is lost or gained by convection or radiation at a rate proportional to the difference between the body surface temperature raised to some power and the surrounding temperature raised to the same power. Chapter 4, Section 4.3, discusses the general details, leading to a relationship involving the normal outward gradient of temperature at the surface, like eqns. (4.3.3) and (4.3.4) of Chapter 4 for example. The objective here is to recalculate the simple problem of Section 6.1, with the more realistic assumption that the slab is immersed in an ambient at 1000 °C and that heat transfer to each face takes place in accordance with a given

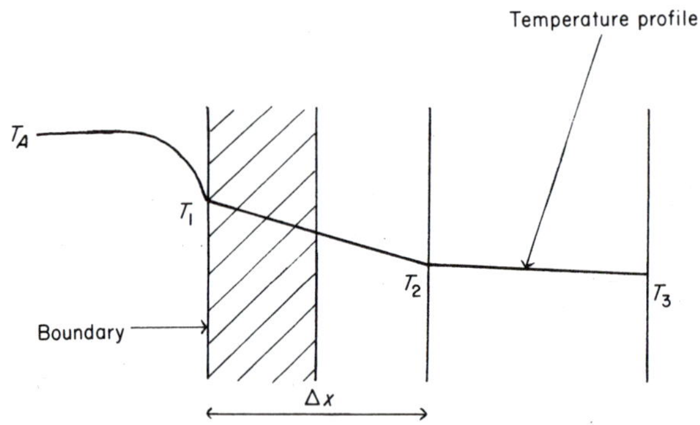

FIG. 6.3. Temperature profile near a convective boundary at the left-hand face of a slab.

heat transfer coefficient h. Figure 4.15 of Chapter 4 illustrates the problem in the vicinity of the left-hand boundary, and Fig. 6.3 illustrates a typical temperature profile in the same vicinity at some later instant in time for our problem, where there is a slight change of notation. Again, six slices across the 210 mm slab are taken. For the left-hand boundary node, the energy balance technique yields the explicit FDE (see eqns. (4.3.14) and (4.3.15) of Chapter 4)

$$T_1' = 2\,\mathrm{Fo}\left[T_2 + \mathrm{Bi}\,T_A + \left(\frac{1}{2\mathrm{Fo}} - 1 - \mathrm{Bi}\right)T_1\right] \tag{6.4.1}$$

where substitution has been made for the Biot number $\mathrm{Bi} = h\Delta x/k$ and the Fourier number

$$\mathrm{Fo} = \alpha \frac{\Delta t}{(\Delta x)^2} = \frac{k}{\rho C_\mathrm{p}} \frac{\Delta t}{(\Delta x)^2}$$

This equation for the left-hand face temperature T_1' must be solved during the forward time-march process along with the usual interior node explicit equations

$$T_i' = \mathrm{Fo}\left[T_{i-1} + T_{i+1} + \left(\frac{1}{\mathrm{Fo}} - 2\right)T_i\right] \quad (i = 2, 3, \ldots) \qquad (6.4.2)$$

which is the generalisation of eqn. (6.1.2) deduced earlier in this chapter. Of course for stability there is a restriction on the value of Fo; eqns. (4.5.2) and (4.5.3) of Chapter 4 reveal that both

$$\mathrm{Fo} \leq 0{\cdot}5 \qquad \text{(interior nodes)} \qquad (6.4.3)$$

and

$$\mathrm{Fo} \leq \frac{1}{2(1 + \mathrm{Bi})} \qquad \text{(boundary nodes)} \qquad (6.4.4)$$

must be obeyed, the more restrictive of these being the latter.

Using the data of Section 6.1, one can compute

$$\alpha = 1{\cdot}017 \times 10^{-5}\,\mathrm{m^2/s}$$

and taking a value of

$$h = 200\,\mathrm{W/m^2\,K}$$

the Biot number is evaluated as

$$\mathrm{Bi} = \frac{200 \times 0{\cdot}21/6}{41} = 0{\cdot}171$$

Hence the stability criterion, from eqn. (6.4.4), is

$$\mathrm{Fo} \leq 0{\cdot}427$$

Taking the limiting value, so as to simplify one of the equations from eqns. (6.4.1) and (6.4.2), the time-step is evaluated as

$$\begin{aligned}\Delta t &= \mathrm{Fo}\frac{(\Delta x)^2}{\alpha} \\ &= \frac{0{\cdot}427 \times (0{\cdot}21/6)^2}{1{\cdot}017 \times 10^{-5}} \\ &= 51{\cdot}4\,\mathrm{s}\end{aligned}$$

In order to obtain the temperature profile after 5 min, linear interpolation is used between time-steps 5 and 6 because the number of steps N_s required is

$$N_s = \frac{300}{51\cdot4} = 5\cdot83$$

Taking the limiting value of Fo and $T_A = 1000$, the equations to be solved, eqns. (6.4.1) and (6.4.2), become

$$T_1' = 0\cdot854T_2 + 145\cdot8$$

$$T_i' = 0\cdot427[T_{i-1} + T_{i+1} + 0\cdot342T_i] \quad (i = 2, 3, \ldots) \qquad (6.4.5)$$

and the computations are shown in Table 6.7, where an average value of T_1 at $t = 0$ is taken between its initial value of 20 °C and the one computed from eqn. (6.4.5). This latter modification is to some extent optional and is discussed in Section 6.1. The linearly interpolated values for conditions after exactly 5 min are also shown in the table. At this instant, the heat flow to each of the two slab faces is

$$hA\,\Delta T = 200 \times 1^2 \times (1000 - 268\cdot9)$$
$$= 146\cdot2\,\text{kW/m}^2$$

The previous computer program, Program 1.2, could equally well compute for this problem as well. The major modification is to the DO 101 and DO 102 loops, which should be replaced by

```
      TP(1) = 0·854★T(2) + 145·8
      DO 101 I = 2,4
      TP(I) = 0·427★(T(I − 1) + T(I + 1) + 0·342★T(I))
  101 CONTINUE
      DO 102 I = 1,4
      T(I) = TP(I)
  102 CONTINUE
```

6.4.2. 2-D Problem with Triangular Mesh

In Section 5.4 of Chapter 5 the steady-state equations were derived for part of a furnace wall problem, in particular points 2 and 8 on the inner convective boundary where triangular segments were required in their vicinity. The task here is to extend those ideas to transient conditions, giving an explicit representation and maintaining the convective boundary condition. Reference is made to Fig. 5.14 of Chapter 5 and attention is now drawn to the cell control volumes (areas) associated with nodes 2 and 8; these are bounded by the regions *abcd*2 and *def*8, which have volumes of $\frac{3}{4}\Delta^2$ and $\frac{1}{2}\Delta^2$, respectively.

TABLE 6.7
STEEL SLAB TEMPERATURE RESPONSE WITH CONVECTIVE BOUNDARIES
(Each time-step represents 51·4 s)

Time-step	T_A	$T_1\,(=T_7)$	$T_2\,(=T_6)$	$T_3\,(=T_5)$	T_4	$T_5\,(=T_3)$
0	1 000	81·5	20	20	20	20
1	1 000	162·9	46·3	20	20	20
2	1 000	185·4	84·9	31·2	20	31·2
3	1 000	218·3	104·9	49·3	29·6	49·3
4	1 000	235·4	129·6	64·6	46·4	64·6
5	1 000	256·5	147·0	84·6	61·9	84·6
6	1 000	271·4	167·1	101·5	81·3	101·5
Linear interpolation for the 5·83 time-step value yields						
5·83	1 000	268·9	163·7	99·1	78·0	99·1

Node 2. The usual heat balance approach, coupled with values of the geometric factors already calculated, gives the equation

$$kF_{12}(T_1 - T_2) + kF_{52}(T_5 - T_2) + kF_{32}(T_3 - T_2) + h(2)^{1/2}\Delta(T_A - T_2) = \rho C_p \tfrac{3}{4}\Delta^2 \frac{2(T_2' - T_2)}{\Delta t}$$

Substituting $F_{12} = F_{52} = F_{32} = 1$, $\text{Bi} = h\Delta/k$ and $\text{Fo} = (k/\rho C_p)(\Delta t/\Delta^2)$ gives the explicit formula:

$$T_2' = \frac{4\text{Fo}}{3}\left[T_1 + T_3 + T_5 + (2)^{1/2}\,\text{Bi}\,T_A + \left(\frac{3}{4\text{Fo}} - 3 - (2)^{1/2}\,\text{Bi}\right)T_2\right] \tag{6.4.6}$$

Node 8. Similar reasoning yields the explicit formula:

$$T_8' = 2\text{Fo}\left[T_3 + T_9 + (2)^{1/2}\,\text{Bi}\,T_A + \left(\frac{1}{2\,\text{Fo}} - 2 - (2)^{1/2}\,\text{Bi}\right)T_8\right] \tag{6.4.7}$$

These two equations must be solved along with other interior and boundary point equations. Stability requirements are invariably more restrictive for convective boundary points than for interior points. For the two such points considered here, the time-step Δt must be restricted according to the more stringent of these two conditions:

$$\text{Fo} \le \frac{3}{[4(3 + (2)^{1/2}\,\text{Bi})]}$$

$$\text{Fo} \le \frac{1}{[2(2 + (2)^{1/2}\,\text{Bi})]}$$

Notice that setting Fo = ∞, that is, taking $\Delta t = \infty$, in eqns. (6.4.6) and (6.4.7) generates the appropriate steady-state equations for the same locations, eqns. (5.4.5) and (5.4.6) of Chapter 5.

6.5. IMPLICIT METHODS FOR TRANSIENT PROBLEMS

Implicit methods for the solution of transient problems overcome the time-step restriction for stability, either partially or totally, as discussed in Chapter 4, Section 4.5. This asset is obtained, however, at the expense of solving a set of simultaneous equations in order to march each step forward. Equations (3.5.7) to (3.5.9) of Chapter 3 are appropriate for fully implicit 3-D, 2-D and 1-D transient problems in the cartesian coordinate system; eqns. (3.5.12) to (3.5.14) are appropriate for weighted-average formulations. In the case of 1-D problems the tridiagonal matrix algorithm (TDMA) is immediately applicable to obtain rapid solution at each step of the march. For 2-D and 3-D problems, however, each point to be updated involves four or more neighbouring point values and no such rapid solution scheme may be used, hence a direct or iterative method is used; Chapter 7 provides a fuller discussion. An alternative technique in these cases is the alternating direction implicit (ADI) method, whereby at each step derivatives in one space direction only are formulated fully implicitly, the others being taken explicitly. Hence a set of rows or columns has to be solved, and the TDMA is applicable for each such row or column. The theory shows (see Carnahan *et al.*, 1969) that these ADI methods are stable provided the direction for implicit representation is changed sequentially for successive forward steps. These techniques always involve the solution of many equations and a computer is invariably used; thus it is more convenient to defer further discussion until Chapter 7, Section 7.5, which follows the introduction and study of suitable computer programs.

BIBLIOGRAPHY

CARNAHAN, B., LUTHER, H. A. & WILKES, J. O. (1969). *Applied Numerical Methods*. Wiley, New York.

DUSINBERRE, G. M. (1961). *Heat Transfer Calculations by Finite Differences*. International Textbook Company, Scranton, Pa.

GERALD, C. F. (1970). *Applied Numerical Analysis*. Addison-Wesley, Reading, Mass.

RAZELOS, P. (1973). Methods of obtaining approximate solutions. Section 4, pp. 4–1 to 4–78, in *Handbook of Heat Transfer*, ed. W. M. Rohsenow and J. P. Hartnett. McGraw-Hill, New York.

SMITH, G. D. (1965). *Numerical Solution of Partial Differential Equations*. Oxford University Press, London.

CHAPTER 7

Computer Solution of Heat Transfer Problems

7.1. OCCURRENCE OF LARGE SYSTEMS OF FDEs

Finite difference simulation of steady-state and transient heat conduction problems (by implicit schemes) leads to systems of many algebraic equations which must be solved simultaneously. Though it may be shown that they have a unique solution, exactly how to compute it is often a difficult task, with considerations of adequacy, accuracy, compatibility and convergence being paramount. Methods of solution may be classed as direct or iterative.

Direct methods, except for a serious build-up of round-off errors when the number of equations is large, obtain the solution directly in a finite number of operations. For linear problems the problem is then finished; but for quasi-linear problems in which, for example, the conductivity is temperature-dependent, the process must be repeated with updated coefficients evaluated based on the newly-computed temperature field. Gauss elimination and variants of it are generally used to effect direct solution efficiently. Matrix inversion by the adjoint cofactor method or Cramer's determinant method are extremely inefficient and should never be used in practice except on, perhaps, fewer than five equations. In any case, the inverse matrix is rarely required and it usually has all non-zero elements even when the given equation set is spare, so that there is an unnecessary core storage requirement. One special case when direct methods are invariably used is on spare systems of equations where the occurrence of non-zero elements follows a particularly simple pattern, like a tridiagonal system where the Gauss elimination direct process takes a particularly simple form known as the tridiagonal matrix algorithm (TDMA) (see Program 2.2). The technique can also be applied to 2-D and 3-D steady-state and transient problems via the alternating direction implicit (ADI) ideas. Such problems, in which each equation contains exactly three unknowns, except the first and last equations which contain two unknowns, result in so-called tridiagonal equation sets, the matrix form of which would comprise a tridiagonal band of coefficients, and these are better solved by direct methods. Such equation sets occur in:

(i) 1-D steady-state problems (see Chapter 5, Section 5.1);
(ii) coarse mesh 2-D steady-state problems, like the problem governed by eqn. (7.1.2); and
(iii) the solution of transient problems by implicit methods (see Chapter 6, Section 6.5).

Iterative methods are generally preferred over direct methods of solution, and when the coefficient matrix is sparse, as is the case in heat transfer problem simulation, they are more rapid. Updating variable coefficients like conductivity can be carried out *en route* easily. They are more economical in core storage requirements and there is no build-up of round-off errors. Indeed, in hand computation they have the distinct advantage that they are self-correcting if an error is made.

Initially, this section is concerned with methods for solving systems of n equations in n unknowns:

$$\begin{aligned} a_{11}T_1 + a_{12}T_2 + \cdots + a_{1n}T_n &= b_1 \\ a_{21}T_1 + a_{22}T_2 + \cdots + a_{2n}T_n &= b_2 \\ \vdots \qquad\qquad\qquad &\quad \vdots \\ a_{n1}T_1 + a_{n2}T_2 + \cdots + a_{nn}T_n &= b_n \end{aligned} \qquad (7.1.1)$$

Finite difference equations of this type have already been derived for several heat transfer situations. As an illustration, the steady-state problem of Section 5.2 of Chapter 5 is reiterated, this time with a finer mesh giving 21 internal node points. It will be seen that many of the coefficients are zero and this fact may be exploited in some solution techniques. Figure 7.1 shows the effect of halving the mesh size for this problem. The coarse mesh, with three unknown node values, is made finer, with 21 unknown node values. The three finite difference equations for the three values in the former case have already been given (eqn. (5.2.4) of Chapter 5); here repeated they are

$$\begin{aligned} -4T_1 + T_2 \qquad\quad &= -400 \\ T_1 - 4T_2 + T_3 &= -400 \\ T_2 - 4T_3 &= -1360 \end{aligned} \qquad (7.1.2)$$

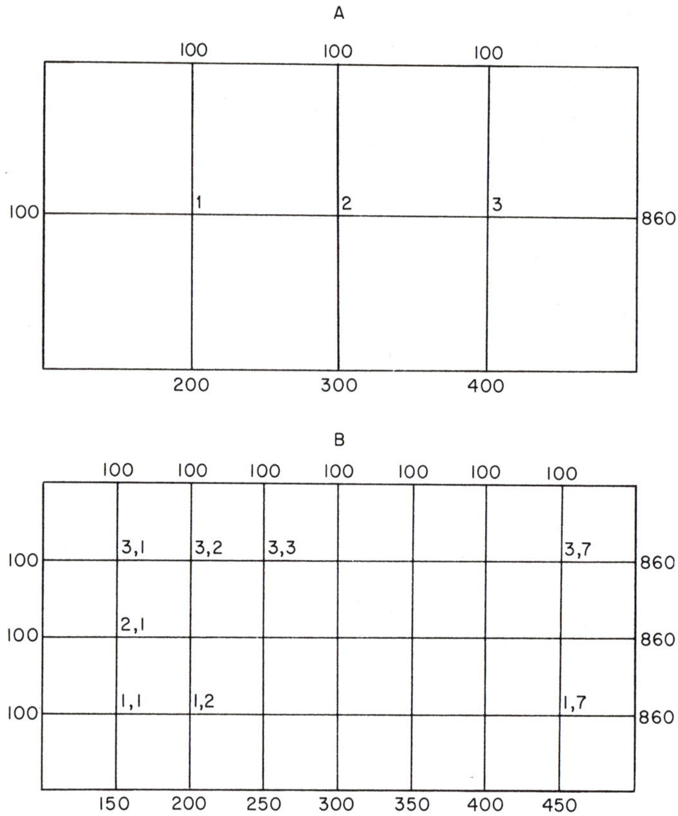

FIG. 7.1. Coarse and fine meshes for a 2-D steady-state rectangular region heat flow problem (see Chapter 5, Section 5.2). A = 3 unknown node values, 3 FDEs (see eqn. (7.1.2)); B = 21 unknown node values, 21 FDEs (see eqn. (7.1.3)).

With the finer mesh, 21 finite difference equations are obtained, one for each of the unknown node point values. As before, but in double suffix notation, the equation for the point (1, 1), for example, is

$$T_{12} + T_{21} - 4T_{11} + 150 + 100 = 0$$

Doing this in order for all 21 points yields the finite difference equation set in matrix form (eqn. (7.1.3)), where the ordering is the usual reading order ((3, 1), (3, 2), ..., (3, 7), (2, 1), (2, 2), ..., (2, 7), (1, 1), (1, 2), ..., (1, 7)).

In general, 2-D steady-state problems, like this one, with a fine mesh,

Temperatures: *Nodes*	T_{31}	T_{32}	T_{33}	T_{34}	T_{35}	T_{36}	T_{37}	T_{21}	T_{22}	T_{23}	T_{24}	T_{25}	T_{26}	T_{27}	T_{11}	T_{12}	T_{13}	T_{14}	T_{15}	T_{16}	T_{17}	
3, 1	−4	+1						+1														= −200
3, 2	+1	−4	+1						+1													= −100
3, 3		+1	−4	+1						+1												= −100
3, 4			+1	−4	+1						+1											= −100
3, 5				+1	−4	+1						+1										= −100
3, 6					+1	−4							+1									= −100
3, 7						+1	−4							+1								= −960
2, 1	+1							−4							+1							= −100
2, 2		+1						+1	−4	+1						+1						= 0
2, 3			+1						+1	−4	+1						+1					= 0
2, 4				+1						+1	−4	+1						+1				= 0
2, 5					+1						+1	−4	+1						+1			= 0
2, 6						+1						+1	−4	+1						+1		= 0
2, 7							+1						+1	−4							+1	= −860
1, 1								+1							−4	+1						= −250
1,2									+1						+1	−4	+1					= −200
1, 3										+1						+1	−4	+1				= −250
1, 4											+1						+1	−4	+1			= −300
1, 5												+1						+1	−4	+1		= −350
1, 6													+1						+1	−4	+1	= −400
1, 7														+1						+1	−4	= −1310

(7.1,3)

generate equation sets with each equation containing up to five unknowns and iterative methods offer the most suitable solution techniques. Notice that the bandwidth here is $15 = 2n + 1$ where $n = 7$; this is because of the row ordering. If column ordering is used it reduces to $7 = 2n + 1$ where $n = 3$. In direct solution methods, a column ordering yielding a narrower band may be exploited in time requirements for the computation.

7.2. DIRECT SOLUTION METHODS

With direct solution methods, a finite number of operations yields the solution but there is a build-up of round-off errors which can swamp the calculated solution. Direct techniques are useful when the number of equations involved is not too large, typically 40 or fewer, or when the equation set is tridiagonal, in which case a set of up to 500 equations is typical.

Elementary elimination may be used to solve an equation set like eqn. (7.1.1). One normally carries out the process methodically, eliminating T_1 from equations 2 to n first, by adding a suitable multiple of the first equation to each of these in turn. This obtains a reduced equivalent set of equations, in which the coefficient of T_1 is zero in all except the first equations. Then T_2 is eliminated from equations 3 to n in a likewise fashion, adding a suitable multiple of equation 2 to each of these in turn. One carries on like this through the rest of the equations, so obtaining an equivalent set of equations in upper triangular form:

$$\begin{aligned} a_{11}T_1 + a_{12}T_2 + \cdots + a_{1n}T_n &= b_1 \\ a'_{22}T_2 + \cdots + a'_{2n}T_n &= b'_2 \\ \vdots \qquad & \quad \vdots \\ a'_{nn}T_n &= b'_n \end{aligned}$$

where primes denote modified coefficients. If at the rth stage the coefficient of T_r in the rth equation is zero, then T_r cannot be eliminated from equations $r + 1$ to n by adding a suitable multiple of equation r to each of these in turn because no such multiple exists. In this case some reordering of equations $r, r + 1, \ldots, n$ can obtain a non-zero coefficient of T_r in the rth equation and the process can continue.

This is always possible, provided the given equation set like eqn. (7.1.1) is solvable, the equations are linearly independent and a unique solution exists, which is equivalent to the matrix of coefficients A:

$$A = \begin{bmatrix} a_{11} & a_{12} & \cdots & a_{1n} \\ a_{21} & a_{22} & \cdots & a_{2n} \\ \vdots & \vdots & & \vdots \\ a_{n1} & a_{n2} & \cdots & a_{nn} \end{bmatrix}$$

having a non-zero determinant:

$$\det A \neq 0$$

and A is then referred to as being invertible or non-singular. When this is not the case, our elimination procedure will break down in its attempt to find the unique solution. This would occur when the system is over- or under-specified, leading to no solution or an infinite number of solutions. However, generally the system possesses n linearly independent equations in the n unknowns $T_1, T_2, \ldots, T_n$ and this complication does not arise.

The last unknown in the system can now be found:

$$T_n = \frac{b'_n}{a'_{nn}}$$

and the rest follow by back-substitution via

$$T_r = \frac{b'_r - \sum_{i=r+1}^{n} a'_{ri} T_i}{a'_{rr}}$$

where $r = n - 1, n - 2, \ldots, 1$ in turn.

To fix these ideas, consider the set of three equations in the three unknowns T_1, T_2 and T_3 for the simple problem of Chapter 5, Section 5.2, which is governed by eqn. (7.1.2). Suitable multiplications and additions reduce the equation set (2), for example, to the upper triangular form

$$\begin{aligned} -4T_1 + \quad T_2 \qquad\quad &= \quad -400 \\ -15T_2 + \;\; 4T_3 &= \; -2\,000 \\ -56T_3 &= -22\,400 \end{aligned}$$

(multiplying the second equation by 4 and adding the first to it, then multiplying the third equation by 15 and adding the second to it), from which back-substitution yields the exact solution:

$$T_1 = 160; \quad T_2 = 240; \quad T_3 = 400$$

Notice that exact arithmetic is used *en route* (no divisions are allowed) and that large numbers occur. With a larger equation set very large numbers occur, with which exact arithmetic, by human or machine, is not possible.

Gauss elimination accomplishes the same task, but to prevent this occurrence of large numbers the diagonal elements are normalised to unity prior to multiplications and additions. Owing to the retention of only a finite number of decimal places, rounding errors occur, but this is the price to pay to prevent large numbers occurring. In any case the computer can retain these only to a certain number of significant figures. The solution by this method is not exact; it is less accurate than the accuracy of each single computation. Thus the final solution obtained suffers from this serious build-up of round-off errors, and in practice only systems of fewer than about 40 equations are solved in this way. Variants of the Gauss elimination procedure include the Gauss–Jordan and Crout reduction or Cholesky methods.

PROGRAM 2.1 (see p. 258)

This program illustrates Gauss elimination. As described, the program computes the upper triangular equivalent equation set, then back-substitution is performed to give a pure diagonal left-hand side. The program shown solves the equation set of three simultaneous equations, eqn. (7.1.2), in fact, but can be used for any system of n equations in n unknowns, occurring typically in boundary value ordinary and elliptic partial differential equation problems. In these latter cases, note that all the coefficients are stored, even though many of them are zero in practice, so the program is wasteful on storage; later programs (notably Programs 2.2 and 4.1) save on storage.

PROGRAM 2.2 (see p. 260)

This program illustrates Gauss elimination when applied to a tridiagonal equation set. The coefficient matrix is tridiagonal and the solution algorithm is known as the tridiagonal matrix algorithm (TDMA). The code shown compresses the matrix so that only these non-zero coefficients and right-hand-side constants are stored, other coefficients being implicitly zero. The computation is very rapid by Gauss elimination since the program automatically exploits this sparseness property.

Using the notation b = below diagonal, d = diagonal, a = above diagonal and c = constants, the notation set is

$$\begin{aligned}
d_1T_1 + a_1T_2 \qquad\qquad\qquad\qquad &= c_1\\
b_2T_1 + d_2T_2 + a_2T_3 \qquad\qquad &= c_2\\
b_3T_2 + d_3T_3 + a_3T_4 &= c_3\\
\ddots \qquad \ddots \qquad \ddots \quad &\\
b_nT_{n-1} + d_nT_n &= c_n
\end{aligned}$$

for which the triangularisation is (primes denote updated values).

$$\text{new } d_i = d_i - \frac{b_i}{d'_{i-1}} \cdot a'_{i-1} \qquad (i = 2, 3, \ldots, n)$$

$$\text{new } c_i = c_i - \frac{b_i}{d'_{i-1}} \cdot c'_{i-1} \qquad (i = 2, 3, \ldots, n)$$

and the back-substitution is

$$T_j = \frac{c'_j - a'_jT_{j+1}}{d_j} \qquad (j = n-1, n-2, \ldots, 1)$$

The program shown solves the five-equation tridiagonal set of five equations:

$$\begin{aligned}
4T_1 - T_2 \qquad\qquad\qquad\qquad &= 0{\cdot}5\\
-T_1 + 4T_2 - T_3 \qquad\qquad\quad &= 1{\cdot}5\\
-T_2 + 4T_3 - T_4 \qquad &= 2{\cdot}5\\
-T_3 + 4T_4 - T_5 &= 3{\cdot}5\\
-T_4 + 4T_5 &= 4{\cdot}5
\end{aligned}$$

where the left-hand side is typical and the right-hand-side constants can be any numbers; those shown are merely for example purposes. The program is especially useful for parabolic partial differential equation problems being solved by an implicit method, for these problems lead to tridiagonal equation sets similar to the one shown, and elliptic partial differential equation problems being solved on a very coarse grid leading to equations like eqn. (7.1.2), or general elliptic problems being solved via a line method where updating is done along a line of values considered at a particular iteration step to be unknown (other points adjacent to this line being considered to be known and given previous iterate values).

7.3. ITERATIVE SOLUTION METHODS

Iterative methods are preferred for large systems of equations typically consisting of from 100 to 1000 and, in some cases, many more equations. There is no build-up of rounding errors as the solution is generated, but in order to guarantee being able to get the solution a dominant diagonal is required. That is, in the previous equation set (eqn. (7.1.1)):

(i) the diagonal coefficient a_{ii} in the ith equation (for each i in turn, $1 < i < n$) must be at least as large in absolute magnitude as the sum of the absolute magnitudes of all the other coefficients on the left of the ith equation; and

(ii) for at least one of the equations the diagonal coefficient must be strictly larger than the others, viz.

$$|a_{ii}| \geq |a_{i,1}| + \cdots + |a_{i,i-1}| + |a_{i,i+1}| + \cdots + |a_{in}| \quad \text{for each } i \quad (1 \leq i \leq n)$$

$$|a_{ii}| \gneqq |a_{i,1}| + \cdots + |a_{i,i-1}| + |a_{i,i+1}| + \cdots + |a_{in}| \quad \text{for at least one } i \quad (1 \leq i \leq n) \qquad (7.3.1)$$

These requirements are usually no problem in practice, for steady-state elliptic problems automatically give rise to equation sets which may be arranged in dominant diagonal form. The method generally consists of starting from an approximate solution (it is often as vague as $T_1 = T_2 = \cdots = T_n = 0$) and repeatedly updating the T values, usually sequentially, many times until repeated updating results in very little change to the T values. Often a convergence criterion such as:

$$\text{maximum fractional change in any } T < 10^{-3}$$

that is,

$$\max_i \left[\frac{(T_i' - T_i)}{T_i}\right] < 10^{-3} \qquad (7.3.2)$$

is used and 10^{-3} is referred to as the tolerance. It is better, however, to ensure that the residuals for all the points have become small, for these measure the departure from exactness at the points in question. We are then said to have converged on to the solution of the equations, which is taken as the latest available set of T values. Typical systems of equations for steady-state problems are eqns. (7.1.2) and (7.1.3). Note that each of these sets already possesses dominant diagonals as defined by eqn. (7.3.1) and so iterative methods should converge to the solution. Updating the T values is done as follows: the dominant diagonal term in each equation is retained on the left and the others are moved to the right. For example, eqn. (7.1.2) is rearranged to

$$\begin{aligned} T_1 &= (400 + T_2)/4 \\ T_2 &= (400 + T_1 + T_3)/4 \\ T_3 &= (1360 + T_2)/4 \end{aligned} \tag{7.3.3}$$

and a typical equation of the equation set (7.1.1) is

$$T_{ij} = \frac{(T_{i+1,j} + T_{i-1,j} + T_{i,j+1} + T_{i,j-1})}{4} \tag{7.3.4}$$

in the notation of Fig. 7.1, where out-of-range values are given the appropriate known boundary temperature.

Each equation is used to update the dominant term now on the left by putting previous iterate or most recent values on the right of the equation and thus calculating a new value for the temperature on the left. Updating techniques differ in complexity and efficiency in their route to the solution, as discussed in Chapter 5, Sections 5.2 and 5.3. The ideas are recalled here with emphasis on handling large systems of such equations and their computational solution. In terms of the refined mesh of Fig. 7.1 and the typical nodal equations, eqn. (7.3.4), the previously described iterative solution techniques are as follows.

7.3.1. Jacobi Iteration

Jacobi (also called Liebmann and simultaneous displacements) is the process which updates sequentially over the entire field of mesh points and a new T_{ij} at the next iteration level (T'_{ij} say) is computed exclusively in terms of previous iterate values at the old iteration level (non-primed T's). The appropriate form of eqn. (7.3.4) is

$$T'_{ij} = \frac{(T_{i+1,j} + T_{i-1,j} + T_{i,j+1} + T_{i,j-1})}{4} \tag{7.3.5}$$

7.3.2. Gauss–Seidel Iteration

Gauss–Seidel (also called successive displacements) is similar to the Jacobi technique but newly computed temperatures are used as soon as they are available. The appropriate form of eqn. (7.3.4) is

$$T'_{ij} = \frac{(T_{i+1,j} + T'_{i-1,j} + T_{i,j+1} + T'_{i,j-1})}{4} \tag{7.3.6}$$

7.3.3. Relaxation

Relaxation is similar to the Gauss–Seidel technique; latest available updated (that is, primed) T's are used as soon as they are available, but this technique differs in that updating is not done sequentially. One decides which temperature to update by scanning the entire field of residuals:

$$R_{ij} = T_{i+1,j} + T_{i-1,j} + T_{i,j+1} + T_{i,j-1} - 4T_{ij} \tag{7.3.7}$$

which measure the departure from exact solution of eqn. (7.3.4) at the mesh point (i,j). One generally works with normalised residuals, which in our case are

$$r_{ij} = \frac{R_{ij}}{4}$$

and the temperature point to update is decided by seeking that point with highest absolute residual r_{ij}. The procedure continues with single point updates, the order being as determined each time by the magnitudes of the residuals calculated *en route*. Each single point update is achieved via the application of

$$T'_{ij} = \frac{(T_{i+1,j} + T_{i-1,j} + T_{i,j+1} + T_{i,j-1})}{4} \tag{7.3.8}$$

or equivalently

$$\begin{aligned} T'_{ij} &= T_{ij} + \frac{R_{ij}}{4} \\ &= T_{ij} + r_{ij} \end{aligned} \tag{7.3.9}$$

Either of these temporarily reduces the residual R_{ij} to zero at the point (i,j); they also alter (usually for the worse) the residual at neighbouring adjacent points $((i+1,j)$, $(i-1,j)$, $(i,j+1)$ and $(i,j-1))$ by amount, in our case, $R_{ij}/16$, or, equivalently, $r_{ij}/4$. For example, the new residual at $(i+1, j)$ becomes

$$R_{i+1,j} = R_{i+1,j} + \frac{R_{ij}}{16}$$

due to the change $\Delta T_{ij} = R_{ij}/4$ being applied at the point (i,j) in order to deplete R_{ij}.

7.3.4. Over-relaxation

Over-relaxation may be used with any of the above three methods. Each point application for one of these processes may be represented by eqns. (7.3.8) or (7.3.9) where newly computed primed T's on the right-hand side are understood in the case of the Gauss–Seidel process, eqn. (7.3.6). As discussed briefly in Chapter 5, Section 5.3, application of these processes zeros the residual at the point in question, usually at the expense of increasing the value of neighbouring residuals previously zeroed. Thus the value of a typical nodal residual oscillates and convergence is relatively slow. This can be avoided if, when updating the temperature at a point, over-relaxation is used, so as not merely to zero the appropriate residual, but to make it take the opposite sign, in expectation that updating nearby nodes on subsequent occasions will have a favourable effect on the over-relaxed residual. In terms of eqn. (7.3.9) over-relaxation uses instead the point-update equation

$$\begin{aligned} T'_{ij} &= T_{ij} + \frac{\omega R_{ij}}{4} \\ &= T_{ij} + \omega r_{ij} \end{aligned} \qquad (7.3.10)$$

which is sometimes written alternatively as

$$\begin{aligned} T'_{ij} &= T_{ij} + \alpha R_{ij} \\ &= T_{ij} + 4\alpha r_{ij} \end{aligned} \qquad (7.3.11)$$

where $\alpha = \omega/r$. Here ω is called the relaxation (or acceleration) parameter and α is called the relaxation factor. For convergence ω and α have upper limits of 2 and 0·5 respectively (compare the idea of avoiding negative coefficients on the right-hand side of the update equation, eqn. (7.3.10) or (7.3.11), with the corresponding 2-D and time-explicit representation and its stability requirement). Values of ω and α equal to 1 and 0·25 respectively represent the usual processes with no over-relaxation. Under-relaxation, with values of ω and α lying between 0 and 1, and 0 and 0·25, respectively, is sometimes used when coefficients are temperature-dependent, in order to suppress oscillations of the coupled temperatures and coefficients. In most cases over-relaxation is used with

$$\begin{aligned} 1 \leq \omega \leq 2 \\ 0{\cdot}25 \leq \alpha \leq 0{\cdot}5 \end{aligned} \qquad (7.3.12)$$

7.3.5. Estimation of Optimum Over-relaxation Parameter

The optimum value (ω_0 or α_0 say) of ω or α for a given problem is not easy to specify. However, when the equation set to be solved has resulted from the simulation of a 2-D steady-state problem, some details are available. Let us restrict the discussion to the Gauss–Seidel iteration process. Optimum ω_0 and α_0 depend on the largest eigenvalue of the iteration matrix associated with the Gauss–Seidel iteration process without over-relaxation. Fairly generally, and certainly in practice, it can be shown, with certain exceptions, that

$$\omega_0 = \frac{2}{[1 + (1 - \lambda)^{1/2}]} \tag{7.3.13}$$

where

$$\lambda = \max|\lambda_i|$$

is the maximum eigenvalue or spectral radius of the Gauss–Seidel iteration matrix. One practical way of estimating λ and hence ω_0 in the case of the Gauss–Seidel iteration process is to compute successive iterate values of the T field when no over-relaxation is used. One then has

$$\begin{aligned} &\text{Iteration 1:} \quad \mathbf{T}^1 = [T_{11}^1, T_{12}^1, \ldots] \\ &\text{Iteration 2:} \quad \mathbf{T}^2 = [T_{11}^2, T_{12}^2, \ldots] \\ &\quad \vdots \qquad\qquad \vdots \qquad\qquad \vdots \end{aligned}$$

Since the errors $\mathbf{e} = [e_{11}, e_{12}, \ldots]$ in $\mathbf{T}$ are changing roughly in a geometric progression, that is,

$$|e_{ij}^{n+1}| \approx \lambda |e_{ij}^n| \tag{7.3.14}$$

where the superscripts represent iteration level, λ can be estimated from

$$\lambda = \lim_{n \to \infty} \left(\frac{\|E_{\text{it}}^{n+1}\|}{\|E_{\text{it}}^n\|} \right) \tag{7.3.15}$$

where the iteration error norm is

$$\|E_{\text{it}}^n\| = \sum_{i,j} |e_{ij}^n|$$

In practice one looks for the value of λ after several iterations with $\omega = 1$, either via eqn. (7.3.15) or simply the use of eqn. (7.3.14) at a particular mesh point, and hence calculates ω_0 and on subsequent iterations uses this amount of over-relaxation.

It may also be shown that when the equations result from simulating steady-state conduction problems, the greatest influence on ω_0 is the mesh spacing Δx, Δy, Δz. For the case of a 2-D problem in a rectangular region with p and q mesh divisions of a square mesh of size Δ on its sides, it can be shown that $\alpha_0 = \omega_0/4$ is the smaller root of the quadratic equation

$$\left(\cos\frac{\pi}{p} + \cos\frac{\pi}{q} \right)^2 \alpha^2 - 4\alpha + 1 = 0 \tag{7.3.16}$$

Also, for this rectangle with p and q mesh divisions, it can be shown that eqn. (7.3.13) is appropriate with

$$\lambda = \cos\frac{\pi}{p} + \cos\frac{\pi}{q}$$

and that for large values of p and q (in practice, greater than 10 is appropriate)

$$\begin{aligned}\omega_0 &= 4\alpha_0\\ &= 4\left[\frac{1}{2} - \frac{\pi}{2\cdot(2)^{1/2}}\left(\frac{1}{p^2}+\frac{1}{q^2}\right)^{1/2}\right]\end{aligned} \qquad (7.3.17)$$

Program 3.1 (see p. 262)

Program 3.1 illustrates Gauss–Seidel iteration as would be applied to a general set of equations like eqn. (7.1.1). It is written so as to allow all coefficients, if necessary, to be non-zero, whereas the equations governing practical problems contain many zeros. It is therefore very wasteful on storage for large sets of equations governing practical problems, all coefficients being read into the A array of dimension N × N after the number of equations to be solved N, the maximum number of iterations ITMAX, and a small value used as a convergence criterion TOL, are read in. The equations given must already be in dominant diagonal form and the initial approximate solution is taken as the zero one, obtained by reading blanks. The standard Gauss–Seidel iteration technique is used, where a typical unknown T_i, dominant in the ith equation of eqn. (7.1.1), is updated via an application of

$$T_i = \frac{[b_i - (a_{i,1}T_1 + a_{i,2}T_2 + \cdots + a_{i,i-1}T_{i-1} + a_{i,i+1}T_{i+1} + \cdots + a_{i,n}T_n)]}{d_{i,i}}$$

or equivalently, in terms of residuals,

$$T_i = T_i + \frac{R_i}{d_{i,i}}$$

where

$$R_i = b_i - (a_{i,1}T_1 + a_{i,2}T_2 + \cdots + a_{i,n}T_n)$$

and in both cases most recent values are used on the right, giving an updated value on the left. The usual prime on an updated T is omitted here and in the program, for replacement is immediate and there is no need to store a whole set of T''s. A single location SAVE is used. Iteration continues until the largest absolute value by which any T value changes is less than TOL, subject to an overriding upper limit of ITMAX iterations. To determine the maximum change, each old T value is saved until after its new value is computed, then the change in its value is compared with DIFF, the largest change of previously computed T values, and if it is greater it replaces the one stored in DIFF. The later program, Program 4.1, is similar but includes several refinements; over-relaxation is allowed and the program is economical on storage.

7.4. COMPUTER SOLUTION OF STEADY-STATE ELLIPTIC PROBLEMS

7.4.1. Laplace Equation

The steady-state Laplace heat flow equation

$$\nabla^2 T = 0$$

has to be represented by a set of FDEs, one such equation being required to hold at each grid point of a mesh (square, rectangular, non-uniform, triangular, or otherwise) covering the region where solution is required; appropriate boundary conditions for T or $\partial T/\partial n$ are also required at all points surrounding this region. To fix the ideas, consider the 2-D version of the above equation:

$$\frac{\partial^2 T}{\partial x^2} + \frac{\partial^2 T}{\partial y^2} = 0$$

which is represented in a uniform square mesh at the point (i,j) by the FDE

$$\frac{1}{\Delta^2}[(T_{i+1,j} - 2T_{ij} + T_{i-1,j}) + (T_{i,j+1} - 2T_{ij} + T_{i,j-1})] = 0$$

where $\Delta = \Delta x = \Delta y$ is the square mesh spacing. This is simply

$$T_{i+1,j} + T_{i-1,j} + T_{i,j+1} + T_{i,j-1} - 4T_{ij} = 0$$

which is the FDE which must hold at each point (i,j) of the region. When all these equations hold to a certain degree of accuracy (that is, for each point (i,j) of the region), one then asserts that the T_{ij} at these discrete points may be taken as good approximations to the true solution, if it could be found and evaluated at the corresponding points.

The point (i,j) may be

(i) on a boundary—in which case the exterior fictitious temperature is eliminated via a central difference approximation to $\partial T/\partial n$ if this gradient condition is appropriate to this point on the boundary, otherwise the equation is replaced by $T_{ij} = K$, where K is the known boundary condition;

(ii) adjacent to a boundary on one or more sides—in which case if the appropriate boundary value is known, then one or more of the first four terms in this equation are known values;

(iii) remote from boundaries—in which case all the first four terms in this equation are unknown as well as T_{ij} itself.

Thus problems of this type reduce to a set of equations, each equation containing at most five unknowns, and there are as many equations as there are unknowns. A typical such problem has been introduced in Section 5.2 of Chapter 5 where 2-D steady-state heat flow in a rectangular region was discussed with known boundary values of T. Coarse and fine square meshes were described at the beginning of this chapter, leading to a set of 3 or 21 FDEs, one equation for each of the mesh points with an unknown temperature. Equation sets (7.1.2) and (7.1.3) resulted; these are typical of steady-state problems and their sparseness is to be noted. Solution of such sets of equations can be via a direct method if the number of equations is less than about 40. Program 2.1 is useful for this. On refining the mesh further to obtain greater accuracy, a greater number of simultaneous equations result which have to be solved (halving the mesh size again for the example in question results in a set of $7 \times 15 = 105$ equations). Direct methods then become unsuitable and iterative methods are appropriate. The previous program, Program 3.1, using Gauss–Seidel relaxation could be used; unfortunately it is written to accept general sets of n equations in n unknowns so that all coefficients are allocated storage spaces (requiring for this example over 10 000 locations). Observing that each equation contains at most five unknowns, this is extremely wasteful in storage and computational time; a program which exploits this sparseness is given below, Program 4.1, where a typical unknown T_{ij} at mesh point (i, j) is updated via an application of eqn. (7.3.6) or, in terms of residuals, eqn. (7.3.9), where in both cases most recent values are used on the right to give an updated value on the left. With over-relaxation, eqn. (7.3.10) is used with an appropriate value for ω or α given by eqn. (7.3.17), although one could use the ideas presented in Section 7.3 for more general problems.

7.4.2. Poisson Equation

If the steady-state problem were accompanied by internal heat generation, requiring the solution of the Poisson equation

$$k \nabla^2 T + H = 0$$

the technique is the same as for the Laplace equation as just described. In 2-D problems a typical finite difference equation for the point (i,j) is

$$\frac{1}{\Delta^2}[(T_{i+1,j} - 2T_{ij} + T_{i-1,j}) + (T_{i,j+1} - 2T_{ij} + T_{i,j-1})] + \left(\frac{H}{k}\right)_{ij} = 0$$

(if H varies with position) where Δ is the square mesh spacing. This rearranges to

$$(T_{i+1,j} + T_{i-1,j} + T_{i,j+1} + T_{i,j-1}) - 4T_{ij} + \Delta^2\left(\frac{H}{k}\right)_{ij} = 0$$

in which T_{ij} is dominant, and this equation is used to update T_{ij} via

$$T'_{ij} = \frac{\left(T_{i+1,j} + T_{i-1,j} + T_{i,j+1} + T_{i,j-1} + \Delta^2\left(\frac{H}{k}\right)_{ij}\right)}{4} \quad (7.4.1)$$

(or a similar equation with some primed T's on the right according to the iteration scheme being used—see Section 7.1). This can be written in the equivalent form, in terms of residuals,

$$T'_{ij} = T_{ij} + \frac{R_{ij}}{4} \quad (7.4.2)$$

where the (i,j)th residual is defined by

$$R_{ij} = T_{i+1,j} + T_{i-1,j} + T_{i,j+1} + T_{i,j-1} - 4T_{ij} + \Delta^2\left(\frac{H}{k}\right)_{ij} \quad (7.4.3)$$

With over-relaxation the obvious extension, eqn. (7.3.10), is used with an appropriate value for ω or α, given by eqn. (7.3.16) or deduced as described in Section 7.3. Notice that if $H = 0$ this equation degenerates to the Laplace equation, so that Program 4.1 below can be applied equally to either case as required.

PROGRAM 4.1 (see p. 266)

This program illustrates the computer solution of 2-D steady-state elliptic problems, solving either the Laplace or Poisson equation by a limited storage Gauss–Seidel iteration technique with over-relaxation (the successive over-relaxation (SOR) method). The program solves on a square mesh with spacing DELX laid on a rectangular region (NWIDE and NHIGH intervals on the sides) with known boundary (Dirichlet) conditions. The accompanying listing solves Poisson's equation

$$k\left(\frac{\partial^2 T}{\partial x^2} + \frac{\partial^2 T}{\partial y^2}\right) + H = 0$$

subject to known $T = 0$ boundary conditions; it can easily be modified to solve the usual Laplace equation with different boundary conditions. Initial data for geometry (NWIDE and NHIGH), conductivity COND and heat generation rate H are read in as data, and initial approximations to the solution are specified as zero in a DO loop. The relaxation factor ALPHA could be specified in the data, but instead its optimum value is calculated by way of eqn. (7.3.17) in the program. The heart of the program is the updating calculation, using eqn. (7.4.2) with slight rearrangement for computational efficiency, enclosed in nested DO loops ensuring a sequential sweep over the mesh points. Iteration continues until convergence such that the maximum change in any T value is less than a tolerance TOL (read in here as 10^{-3}) with an overriding upper limit of 50 iterations. Finally, the heat flow by the conducting-rod analogy (Section 5.2 of Chapter 5) is computed and compared with the internal heat generation; the discrepancy is because of the use of finite difference approximations and the use of a finite convergence criterion TOL. It will reduce as the mesh is refined and/or TOL is reduced.

7.4.3. Other Details

(i) *Derivative boundary conditions*. These conditions are also referred to as Neumann conditions or convective boundaries at which normal gradients are known or implied via a simple law. If at some point on a boundary the temperature is not known, one must solve an extra algebraic finite difference equation for the boundary mesh point in question. This equation involves a fictitious exterior point at which the temperature is also unknown—it is not equal to the ambient temperature—and this is eliminated by writing the known or implied normal gradient at the boundary point in terms of central differences, thus maintaining $0(\Delta^2)$ accuracy, the same as at internal points of a uniform mesh when using the usual five-point formula. Figure 7.2 illustrates this for the case of an eastern boundary of a 2-D rectangular body covered by a square mesh. The ideas now presented have been met before (see Section 4.3 of Chapter 4); here we restrict the discussion to the PDE replacement method but remark that the same result is obtained if the energy balance technique is used instead. We need to obtain a FDE for T_B, the unknown boundary temperature, in terms of other unknowns already in the system. The usual FDE is assumed to hold at the boundary point B, which involves temperatures at N, S, F and W also. For the Laplace equation in a square mesh this is simply

$$T_N + T_S + T_F + T_W - 4T_B = 0 \tag{7.4.4}$$

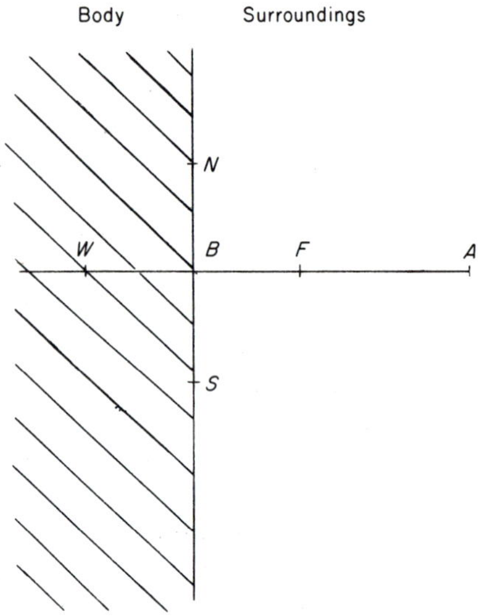

FIG. 7.2. Mesh points used with derivative boundary conditions at an eastern boundary of a 2-D body. W = first interior node, B = surface node, F = fictitious point, A = ambient conditions.

As these temperatures are all in the system of equations already, except the one at the fictitious point F, another equation is now required from which T_F can be eliminated. Two cases may arise, the second one of which is more realistic:

(a) If

$$\left.\frac{\partial T}{\partial x}\right|_B = C \quad \text{(known value)}$$

then, using central differences for the derivative at B,

$$\frac{T_F - T_W}{2\Delta} = C \tag{7.4.5}$$

where Δ is the uniform square grid spacing, and so

$$T_F = T_W + 2\Delta C$$

from which T_F can now be eliminated from eqn. (7.4.4) to yield an equation connecting B with its three neighbours:

$$T_N + T_S + 2T_W - 4T_B = -2\Delta C$$

(b) If $\partial T/\partial x|_B$ is implied through knowledge of rate of heat loss by convection, then

$$k\frac{\partial T}{\partial x}\bigg|_B = -h(T_B - T_A)$$

is appropriate, from which, using central differences for the derivative again,

$$\frac{T_F - T_W}{2\Delta} = -\frac{h}{k}(T_B - T_A)$$

and so

$$T_F = T_W - 2\Delta\frac{h}{k}(T_B - T_A)$$

Often $\Delta h/k$ is replaced by Bi, the Biot number, in this expression. Since T_A is the known ambient temperature ($\neq T_F$), this equation now serves to allow T_F to be eliminated from eqn. (7.4.4). The resulting equation

$$T_N + T_S + 2T_W - (4 + 2\text{Bi})T_B = -2\text{Bi}T_A$$

could also have been obtained directly from the heat balance approach, as in Section 4.3 of Chapter 4.

(ii) *Mesh systems.* Overlaying a rectangular region with a rectangular mesh is simple. Usually on deduces Δx from the size $x = 0$ to $x = \text{XL}$ of the region covered by the body and the number N of grid intervals required. The Fortran code

```
      XL = 5.0
      N = 10
      NP1 = N + 1
      DX = XL/N
      X(1) = 0.0
      DO 100 I = 2, NP1
      X(I) = X(I - 1) + DX
  100 CONTINUE
```

fixes all the x values of the $N + 1$ grid lines, including the first and the last. If Dirichlet boundary conditions are appropriate at $x = 0$ and $x = \text{XL}$ this grid is all right; but if Neumann conditions are appropriate at, for example, $x = 0$, it is sometimes convenient to make X(1) the fictitious point and

X(2) = 0. Then the code

```
      XL = 5.0
      N = 10
      NP1 = N + 1
      NP2 = N + 2
      DX = XL/N
      X(1) = -DX
      DO 100 I = 2, NP2
      X(I) = X(I - 1) + DX
  100 CONTINUE
```

is appropriate. Similarly one can arrange for an extra row of mesh points to be at x = XL + DX, just to the right of the right-hand boundary. The grid lines in the y-direction can be set in a similar fashion.

When a problem is known or expected to possess a solution which shows great activity in certain regions, it is convenient and useful, from the point of view of accuracy and efficiency, to arrange for a gradually expanding or contracting grid in either the x- or y-directions. This can be done in several ways, but the following is perhaps conceptually the easiest to follow. Notice that it is now required to store the DX array, as these values are used in the FDE for each point (I, J) and they differ according to the value of I. Suppose the x grid lines are required to contract at an expansion ratio EPSX = 0·9 so that the mesh is finer nearer the right-hand boundary. One first computes a grid with the ratio EPSX giving the required contraction between adjacent grid intervals. This is done with an approximate DXAPP and the last point X(NP1) will not correspond with x = XL. The appropriate multiple XL/X(NP1) then multiplies each X(I) value to fit the body exactly and retain the expansion ratio specified. Incidentally, this ratio need not be constant throughout the x-range and can be varied as one wishes. The following Fortran coding is recommended:

```
      XL = 5.0
      EPSX = 0.9
      N = 10
      NP1 = N + 1
      DXAPP = XL/N
      X(1) = 0.0
      DO 100 I = 2, NP1
      X(I) = X(I - 1) + DXAPP
      DXAPP = DXAPP*EPSX
  100 CONTINUE
```

```
      FACTOR = XL/X (NP1)
      DO 101 I = 2, NP1
      X(I) = X(I)★FACTOR
      DX(I) = X(I) – X(I – 1)
101   CONTINUE
```

Note that an array for DX is now filled—it is used in the non-uniform mesh formulae which must be employed at all the points. One would then set up locations for coupling coefficients, as in Program 2.5 of Chapter 5. Although it may be anticipated that one loses an order of accuracy (non-uniform versus uniform grid) in the discretisation error, this is amply recompensed by expending effort in regions where it is required; and overall greater accuracy for a given computer time is obtained. One should not, however, let EPSX depart too far from unity: $0{\cdot}8 \leq \text{EPSX} \leq 1{\cdot}25$ is recommended.

(iii) *Overlapping boundary grid systems.* A unified approach to handling both Dirichlet and Neumann boundary conditions alike is simple and attractive. This can be achieved by having the grid system overlap the boundaries, exactly so that the last grid interval is bisected by the boundary. Obviously this is only possible when the body geometry lies parallel to grid lines; that is, a cylindrical body in the case of cylindrical polar coordinates, or a rectangular body in the case of cartesian coordinates. Here the discussion is restricted to a 2-D cartesian rectangular region, as shown in Fig. 7.3, with an overlapping uniform rectangular grid. An expanding grid can also be incorporated with a little more complication. The grid system

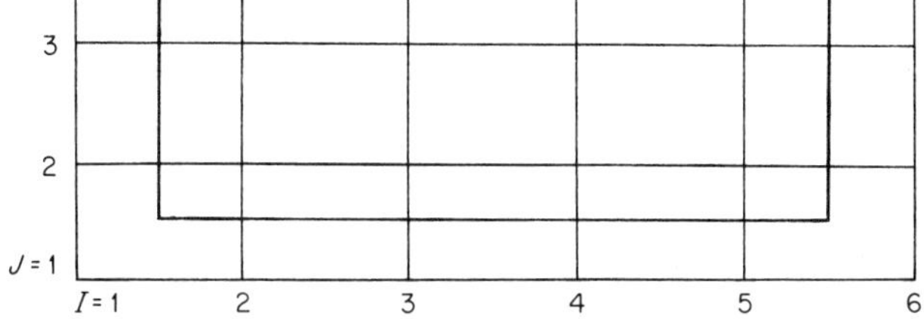

FIG. 7.3. Overlapping grid used for a 2-D rectangular region.

shown has the outstanding property that body temperature update is restricted to the points with $2 \leq I \leq 5$ and $2 \leq J \leq 4$, regardless of the types of boundary conditions. Prior to internal field iteration or time-level advance, the external temperatures at points with at least one of I equal to 1 or 6, or J equal to 1 or 5, are updated. This exterior point setting differs according to the type of boundary condition as follows.

If a particular boundary is known to be adiabatic, so that $\partial T/\partial n = 0$ there, it is more accurate to employ an overlapping grid so that the boundary lies half-way between the grid lines. The first internal line of points where temperatures are to be found are along X(2). For these one assumes that the usual equation holds but that influence from the left-hand neighbour at X(1) is zero. This can be obtained by setting to zero the appropriate coefficient (AE(2, J) = 0 in this case for each J) or alternatively setting the external temperatures at X(1) equal to the current one at X(2) prior to each update sweep of the internal temperature field. Minor modifications allow convective boundaries to be handled similarly, when a central difference approximation to the normal gradient is appropriate.

In order to retain this same overlapping grid system at boundaries where temperatures are known, thus retaining a uniform approach on all the boundaries, it is necessary, prior to each update sweep of the internal points, to set the immediate exterior temperature. This is done by simple linear extrapolation from the current internal temperature and the known boundary temperature. For example, if the boundary temperature is T_B and the current first internal temperature is T_2, then the extrapolated temperature T_1 is

$$\begin{aligned} T_1 &= T_B + (T_B - T_2) \\ &= 2T_B - T_2 \end{aligned}$$

In particular, if $T_B = 0$ then the external temperature setting is

$$T_1 = -T_2$$

which intuitively gives $T_B = 0$ by linearising between T_1 and T_2.

(iv) *Accuracy of solutions.* Any of the techniques already mentioned in Chapter 4, Section 4.4, are applicable to improve the accuracy. These include use of a finer mesh, extrapolating to zero mesh size from values at a point obtained with two mesh sizes (one-half the size of the other), the deferred-correction method and use of a more accurate finite difference representation (using nine or more points in each FDE as opposed to the usual number of five). Since these have been discussed at length already, there is no need to elaborate further here.

7.5. COMPUTER SOLUTION OF TRANSIENT PARABOLIC PROBLEMS

7.5.1. General

Transient heat flow problems of the form

$$\rho C_p \frac{\partial T}{\partial t} = k\,\nabla^2 T + H$$

where the term on the left-hand side is non-zero, may be solved subject to initial and edge boundary conditions. For example, in one space dimension the equation for $T = T(x, t)$:

$$\frac{\partial T}{\partial t} = \frac{k}{\rho C_p}\frac{\partial^2 T}{\partial x^2} = \alpha\frac{\partial^2 T}{\partial x^2} \tag{7.5.1}$$

(where α = thermal diffusivity) may be solved for $0 \le x \le L$ and $t > 0$ subject to initial conditions ($t = 0$, $0 \le x \le L$):

$$T(x,t)|_{t=0} = T(x,0) = f(x) \quad \text{(known)}$$

and edge boundary conditions ($t > 0$, $x = 0$ and $x = L$):

$$T(0,t) = c_1 \quad \text{and} \quad T(L,t) = c_2$$

where c_1 and c_2 are known constants or functions of time. With this example, several solution methods will now be discussed. Simple extension of the work presented can account for the effects of derivative boundary conditions, of the type found with convective boundaries for example. The incorporation of these has already been discussed in Section 7.4 and is not described in detail here. However, Program 5.2 incorporates these additional complications.

7.5.2. Explicit Solution Method

The explicit method is an especially simple method which suffers from a time-step Δt restriction for stability. A rectangular mesh of size Δx and Δt covers the region where the solution is required. The method consists of beginning with the known initial T values and marching forward in the t-direction, at each step going from known values at t_j (the jth step) to values at t_{j+1} (the $j+1$th step). For this step to $(i, j+1)$, point (x_i, t_{j+1}), the forward difference operator

$$\frac{\partial T}{\partial t} = \frac{T_{i,j+1} - T_{i,j}}{\Delta t} + 0(\Delta t)$$

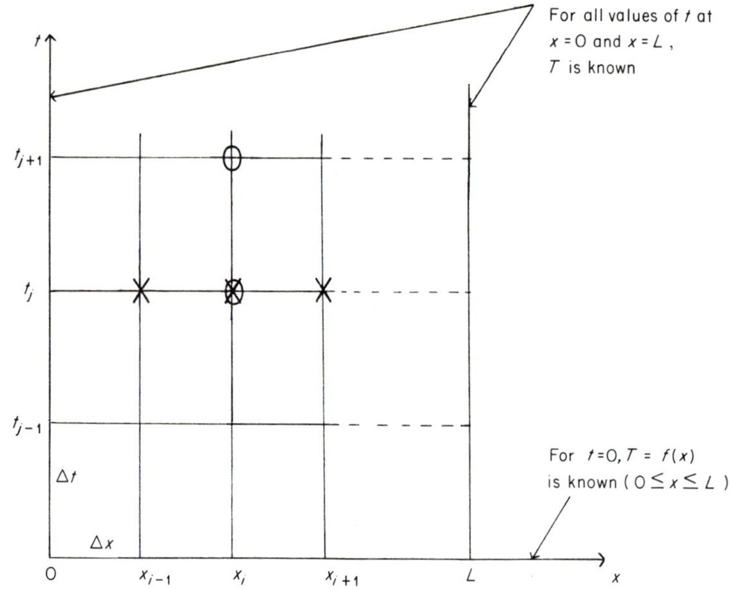

FIG. 7.4. Explicit finite difference simulation for 1-D transient problems (x = points used for space derivative, 0 = points used for time derivative).

is used for the time derivative, but backward values at the t_j time-step are used (in the explicit method) for the second-order space derivative:

$$\frac{\partial^2 T}{\partial x^2} = \frac{T_{i+1,j} - 2T_{i,j} + T_{i-1,j}}{(\Delta x)^2} + 0(\Delta x)^2$$

Thus eqn. (7.5.1) is replaced by

$$T_{i,j+1} = \alpha \frac{\Delta t}{(\Delta x)^2}(T_{i+1,j} + T_{i-1,j}) + \left(1 - 2\alpha \frac{\Delta t}{(\Delta x)^2}\right) T_{i,j} \quad (7.5.2)$$

and this is used at each step for all the points across the region from $x = 0$ to $x = L$. Figure 7.4 illustrates this explicit solution technique. For a given Δx subdivision, if Δt is so chosen that

$$\text{Fo} = \text{Fourier number} = \frac{\Delta t}{(\Delta x)^2} = \frac{k}{\rho C_p} \frac{\Delta t}{(\Delta x)^2} = 0{\cdot}5$$

the equation simplifies to

$$T_{i,j+1} = 0{\cdot}5(T_{i+1,j} + T_{i-1,j}) \quad (7.5.3)$$

which is known as the Binder–Schmidt explicit method. For stability (see Chapter 4, Section 4.5) it may be shown that this value of Δt is the largest that is allowed when edge conditions are known; this is equivalent to the previously discussed requirement with transient problems of avoiding negative coefficients on the right, or equivalently of keeping $\text{Fo} \leq 0{\cdot}5$.

The Binder–Schmidt explicit method consists of a graphic method of representing successive time-steps to the solution of eqn. (7.5.1) via the successive use of the FDEs, eqn. (7.5.3). The ordinates represent the values of T attained at particular abscissae as time t progresses. As an example, it allows the end boundary condition of a rod, suddenly heated at one end, to propagate along the rod as time progresses.

PROGRAM 5.1 (see p. 271)

The explicit method for a 1-D transient (parabolic) problem with known (Dirichlet) edge conditions is illustrated in Program 5.1. Both ends of a rod of length $L = 10$ cm are held for $t > 0$ at constant temperatures: 0 °C on the left and 100 °C on the right. Initially ($t = 0$) $T = 20$ °C at all points along the rod but, for accuracy reasons, the end temperatures for $t = 0$ are taken as averages of their initial and subsequent ones, that is, 10 °C on the left and 60 °C on the right. With these boundary conditions the governing one space dimension and time transient equation

$$\frac{\partial T}{\partial t} = \alpha \frac{\partial^2 T}{\partial x^2}$$

is solved by use of the fundamental equation, eqn. (7.5.2), and this is used at each internal point, successively across the rod at each time-step. After each row of temperatures is computed at a certain time-step they are printed out. It is to be noted that as time proceeds the temperatures calculated tend to the steady-state solution, which could be obtained (for this simple example) by direct solution of the one-dimensional Laplace equation

$$\frac{\partial^2 T}{\partial x^2} = 0$$

which, fitting the end conditions, is $T = 100x/L$. Hence T tends to 20°, 40°, 60° and 80 °C at the four internal points used in the example calculation.

7.5.3. Implicit Solution Methods

(i) *Fully implicit method.* More complicated to apply than the explicit method, this method has the advantage of being stable for any Δt (that is, for all choices of Fo) at the expense of a larger truncation error at every step and accuracy suffers slightly, this accuracy depending on Δt: the smaller Δt, the more accurate the computations. Thus reasonably small Δt is required for reasonable accuracy. The method represents the second-order space derivative in terms of forward unknown values at the time-step t_{j+1} via

$$\frac{\partial^2 T}{\partial x^2} = \frac{T_{i+1,j+1} - 2T_{i,j+1} + T_{i-1,j+1}}{(\Delta x)^2} + 0(\Delta x)^2$$

so that the finite difference equation for calculating $T_{i,j+1}$ is

$$-\text{Fo}\, T_{i-1,j+1} + (1 + 2\text{Fo})T_{i,j+1} - \text{Fo}\, T_{i+1,j+1} = T_{i,j}$$

In the implicit method this is calculated at the time-step t_j, that is, it contains three unknown values. For each step of the march this equation is written once for each of the points across the region from $x = 0$ to $x = L$ (see Fig. 7.5). Thus for a single step to a new time-line a system of equations has to be solved. Each equation contains three unknowns, except the equation for the first and last point at which solution is sought which contains only two unknowns, and the system is tridiagonal, when the equations are written down sequentially, for which the previous Program 2.2 is readily applicable.

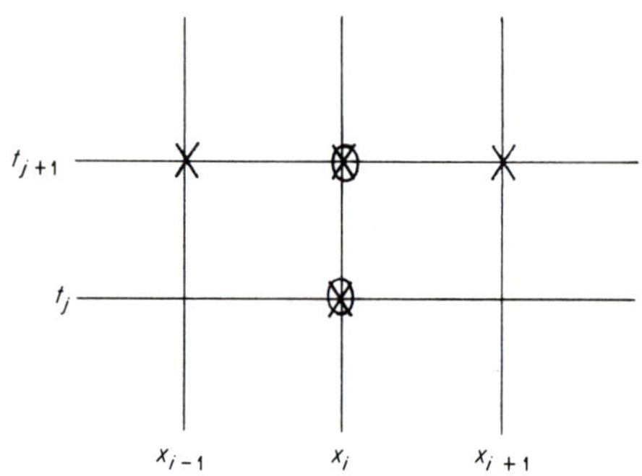

FIG. 7.5. Fully implicit finite difference simulation for 1-D transient problems (x = points used for space derivative, 0 = points used for time derivative).

(ii) *General implicit method.* The fully implicit method, though stable for any Δt, is not as accurate for a given amount of computational work as the more general implicit method obtained by evaluating the second-order space derivative at an intermediate point between t_j and t_{j+1}; a weighted average of its values at t_j and t_{j+1} is used:

$$\frac{\partial^2 T}{\partial x^2} = \lambda \left.\frac{\partial^2 T}{\partial x^2}\right|_{j+1} + (1 - \lambda)\left.\frac{\partial^2 T}{\partial x^2}\right|_{j}$$

where $0 < \lambda \leq 1$. For a given Δt and Δx, there is a progressive decrease in accuracy as λ increases to 1.

The finite difference equation for evaluating $T_{i,j+1}$ becomes

$$-\mathrm{Fo}\,\lambda T_{i-1,j+1} + (1 + 2\mathrm{Fo}\,\lambda)T_{i,j+1} - \mathrm{Fo}\,\lambda T_{i+1,j+1}.$$
$$= \mathrm{Fo}(1-\lambda)T_{i-1,j} + [1 - 2\mathrm{Fo}(1-\lambda)]T_{i,j} + \mathrm{Fo}(1-\lambda)T_{i+1,j} \quad (7.5.4)$$

Putting $\lambda = 0$ gives the explicit form; $\lambda = 1$ gives the fully implicit form. We can choose Fo (by deciding on Δt for a given Δx subdivision) and also choose the weighting value λ. The question of what values are best naturally occurs. A schematic of the position is given below:

Implicit $\lambda = 1$ C–N $\lambda = 0{\cdot}5$	Stable for all Fo or Δt	High truncation error Restrict Fo (or Δt) for accuracy Low truncation error
$0 \leq \lambda \leq 0{\cdot}5$ Explicit $\lambda = 0$	Stable only if Fo (or Δt) is restricted by $\mathrm{Fo} \leq 0{\cdot}5/(1-2\lambda)$	Low accumulated truncation error

(iii) *Crank–Nicolson method.* Each of these implicit methods involves solution of a set of equations, each with three unknowns except the first and last with exactly two unknowns. Program 2.2 is used in order to march one time-step forward. The choice for λ is generally up to the user. If there are other complications in the equation being solved, for example variable conductivity on a heat generation term, one would opt for a more fully implicit representation obtained by taking λ close to 1, for this choice generates the most stable FDE albeit at the expense of local truncation error and accuracy. All simple systems with $0{\cdot}5 \leq \lambda \leq 1$ are stable for all Fo or Δt. If the problem is relatively straightforward and the stability restriction on Fo (or Δt) is not too restrictive, a choice of λ nearer to 0 is appropriate. If $0 \leq \lambda < 0{\cdot}5$, the FDEs are stable only if

$$\mathrm{Fo} \leq \frac{0{\cdot}5}{(1-2\lambda)}$$

and these representations then possess low accumulated truncation error and hence good accuracy when used to compute to a given time-position. In particular, if the time-step length is chosen so that

$$\mathrm{Fo} = \frac{1}{[6(1-2\lambda)]}$$

then the truncation error is very low.

However, the choice $\lambda = 0{\cdot}5$ gives the least accumulated truncation error of the always stable systems. Then the form of the growing FDE, eqn. (7.5.4), becomes

$$\frac{-\mathrm{Fo}}{2}T_{i-1,j+1} + (1 + \mathrm{Fo})T_{i,j+1} - \frac{\mathrm{Fo}}{2}T_{i+1,j+1} = \frac{\mathrm{Fo}}{2}T_{i-1,j} + (1 - \mathrm{Fo})T_{i,j} + \frac{\mathrm{Fo}}{2}T_{i+1,j} \qquad (7.5.5)$$

and the application of this is known as the Crank–Nicolson method. The form of this depends on the choice for Fo (or Δt), and though the formula is stable for all choices, the choice Fo = 1 yields the particularly simple form of the Crank–Nicolson formulation which is often the form to be recommended:

$$-T_{i-1,j+1} + 4T_{i,j+1} - T_{i+1,j+1} = T_{i-1,j} + T_{i+1,j} \qquad (7.5.6)$$

PROGRAM 5.2 (see p. 275)

Program 5.2 illustrates the computer solution of a 1-D transient (parabolic) problem using the implicit Crank–Nicolson method employing eqn. (7.5.6) and with the additional complexity of derivative (Neumann) edge boundary conditions properly accounted for. These derivative end conditions representing convection heat losses are permitted in the form

$$\frac{\partial T}{\partial x} = A(T - T_A)$$

where A is a constant (equal to h/k on the left and $-h/k$ on the right), T is a rod end temperature and T_A is the ambient temperature. Initial temperatures along the rod are taken linearly from 50° to 100°C and computations ensue until the midpoint temperature falls to 60°C. To do this, the program generates the coefficients of the set of simultaneous equations, eqn. (7.5.6), that are to be solved in order to advance one time-step, Δt being calculated from read-in values of Δx and Fo = 1. These equations are, using the method of fictitious points just beyond the ends:

At the left-hand end:

$$-(4 + 2A\,\Delta x)T_{1,j+1} + 2T_{2,j+1} = -2T_{2,j} + 2A\,\Delta x(T_{1,j} - 2T_A)$$

At interior points:

$$T_{i-1,j+1} - 4T_{i,j+1} + T_{i+1,j+1} = -T_{i-1,j} - T_{i+1,j}$$

At the right-hand end:

$$2T_{n-1,j+1} - (4 + 2A\,\Delta x)T_{n,j+1} = -2T_{n-1,j} + 2A\,\Delta x(T_{n,j} - 2T_A)$$

These equations form a tridiagonal equation set which is solved via a subroutine incorporating the method of Program 2.2 (tridiagonal system by Gauss elimination). Each call to this subroutine results in temperatures at the new time-step being computed. Printing is arranged to occur about every 100 s time-intervals. A comparison of this approach with the fully implicit approach is given in Section 8.7 of Chapter 8.

7.5.4. Transient Problems in Two or More Space Dimensions

Consider a 2-D transient problem. The explicit finite difference formula associated with

$$\frac{\partial T}{\partial t} = \frac{k}{\rho C_{\mathrm{p}}}\left(\frac{\partial^2 T}{\partial x^2} + \frac{\partial^2 T}{\partial y^2}\right)$$

where

$$T = T(x, y, t)$$

$$\mathrm{Fo} = \frac{k}{\rho C_{\mathrm{p}}} \cdot \frac{\Delta t}{(\Delta x)^2}$$

is, taking a square x-y grid of size Δx,

$$T_{i,j,k+1} = \mathrm{Fo}(T_{i+1,j,k} + T_{i-1,j,k} + T_{i,j+1,k} + T_{i,j-1,k}) + (1 - 4\mathrm{Fo})T_{i,j,k}$$

and is stable if $\mathrm{Fo} \leq 0{\cdot}25$ in the case of known edge boundary conditions. There is an obvious extension to three space dimensions for which the stability criterion is $\mathrm{Fo} \leq 1/6$. To remove the time-step restriction for stability, one looks for an implicit method. Any of the weighted implicit formulae for $T_{i,j,k+1}$ involves four other values at the t_{k+1} time-level which are also unknown. One has to solve the whole set of equations for one time-step movement; each equation has five unknowns and a dominant diagonal, and a Gauss–Seidel iterative solution technique (Program 3.1 inefficient in storage or preferably Program 4.1 efficient in storage) has to be carried out for each time-step. One has therefore to solve an elliptic problem by relaxation for each of the time-steps of the march.

A better alternative is the alternating direction implicit (ADI) method which is stable for all Δt but requires merely the solution of several sets of tridiagonal equation sets for each time-step movement. The method consists of representing one of the second-order space derivatives fully implicitly (for rows, say) and the other explicitly (for columns, say), then solving to go to the new time-level t_{j+1}.

This involves solving for each row in turn, a tridiagonal set being inverted for each row in order to make this time advance. When all the rows have been dealt with sequentially like this, one can delete the t_j-level and prepare for the t_{j+2} time-level from the t_{j+1} values; the technique is similar but the implicit–explicit representation is reversed from that of the previous step. For the case just described this new step would be accomplished by representing the columns implicitly and the rows explicitly this time, and then solving a tridiagonal set for each column in turn to get values at time-level t_{j+2}. This alternation of directions to represent explicitly or implicitly is necessary in order to ensure stability of the computations, for it can be shown that not alternating the directions results in an unstable process. For these types of problems, this ADI method is the one to be recommended, on account of efficiency, accuracy and stability. It can also be used if one is interested only in the eventual steady-state solution, to be computed as described above, or alternatively if one is iteratively solving the associated steady-state equation set. In this latter case, one application of the ADI processes advances the computations not to a new time-level but to a new iteration-level. Convergence is often more rapid than by the more usual Gauss–Seidel process as illustrated in Program 4.1. Readers interested in the mechanism of these ADI methods are referred to the literature.

BIBLIOGRAPHY

CARNAHAN, B., LUTHER, H. A. & WILKES, J. O. (1969). *Applied Numerical Methods*. Wiley, New York.

FORSYTHE, G. E. & WASOW, W. R. (1960). *Finite-Difference Methods for Partial Differential Equations*. Wiley, New York.

FOX, L. (1964). *Introduction to Numerical Linear Algebra*. Oxford University Press, London.

GERALD, C. F. (1970). *Applied Numerical Analysis*. Addison-Wesley, Reading, Mass.

GOULT, R. J., HOSKINS, R. F., MILNER, J. A. & PRATT, M. J. (1974). *Computational Methods in Linear Algebra*. Stanley Thornes Ltd, London.

HORNBECK, R. W. (1974). *Numerical Marching Techniques for Fluid Flow with Heat Transfer*. NASA SP-297, NASA, Washington, DC.

NATIONAL PHYSICAL LABORATORY (1961). *Modern Computing Methods*, 2nd ed. Notes on Applied Science Series No. 16, HMSO, London.

POLACHEK, H. (1965). Numerical analysis. In: *Research Frontiers in Fluid Dynamics*, ed. R. J. Seeger and G. Temple. Wiley, New York.

SCHEID, F. (1968). *Numerical Analysis*. Schaum Outline Series, McGraw-Hill, New York.

SMITH, G. D. (1965). *Numerical Solution of Partial Differential Equations*. Oxford University Press, London.

TODD, J. (ed.) (1962). *Survey of Numerical Analysis*. McGraw-Hill, New York.

VARGA, R. S. (1962). *Matrix Iterative Analysis.* Prentice-Hall, Englewood Cliffs, NJ.

VON ROSENBERG, D. U. (1969). *Methods for the Numerical Solution of Partial Differential Equations.* Elsevier, Amsterdam.

WALSH, J. (1966). Elliptic and parabolic equations. Chapter 5, pp. 99–116, in *Numerical Analysis: An Introduction*, ed. J. Walsh. Academic Press, London.

YOUNG, D. (1966). The numerical solution of elliptic and parabolic partial differential equations. In: *A Survey of Numerical Analysis*, ed. J. Todd, McGraw-Hill, New York.

CHAPTER 8

Miscellaneous Problems

The purpose of this chapter is to illustrate how all the techniques and procedures previously expounded in this book are integrated to solve realistic problems. The approach to each problem is the same, comprising:

(i) a problem statement in which the basic nature of the problem is explained and rationalised to a 2-D transient problem, for example, or a 2-D steady-state problem with variable thermal properties, etc.;
(ii) a discussion of the type and size of grid to be employed;
(iii) the development of the finite difference equations for every typical node point and the presentation of these equations in matrix or other form;
(iv) the choice of a solution procedure;
(v) the computed results and a discussion of the results in terms of the initial problem statement.

Table 8.1 summarises the problems covered in this chapter.

8.1. STEADY-STATE HEAT FLOW THROUGH A FURNACE WALL

8.1.1. Problem Statement

Figure 8.1 shows the plan view of the cross-section of a small furnace, the inside wall of which is subjected to a temperature of 1200 °C; the outer wall is cooled by convective heat transfer. The variation of temperature with

TABLE 8.1
SUMMARY OF PROBLEMS IN CHAPTER 8

Problem	*Problem Description*	*Coordinate System*	*Boundary Conditions*	*Derivation of FDE*	*Solution Procedure*	*Other Special Features*
8.1	2-D s/s	Cartesian	Dirichlet and Convective	Energy balance, explicit	Gaussian elimination	
8.2	2-D s/s + H	Cartesian	Dirichlet	Energy balance, explicit	Direct TDMA	Effect of grid size
8:3	2-D Tr. + H	Cartesian	Convective	Energy balance, explicit	Iterative	
8:4	1-D Tr. + H	Cylindrical	Convective	Direct replacement, explicit	Iterative	Isolated internal heat generation
8.5	2-D Tr. + s/s	Triangular	Convective	Energy balance, explicit	Gaussian elimination and iterative	
8.6	1-D Tr.	Cartesian	Dirichlet	Energy balance, explicit	Iterative	Change of phase
8.7	1-D Tr.	Cartesian	Dirichlet	Energy balance, implicit and explicit	Iterative	Comparison of explicit/implicit
8.8	2-D Tr.	Cartesian	Convective	Direct replacement, implicit and explicit	Direct TDMA	ADI method
8.9	1-D Tr. + s/s	Cartesian	Dirichlet and Convective and Radiative	Energy balance, explicit	Relaxation	Composite structure, Variable thermal properties
8.10	3-D s/s	Cartesian	Convective	Energy balance, explicit	Relaxation	

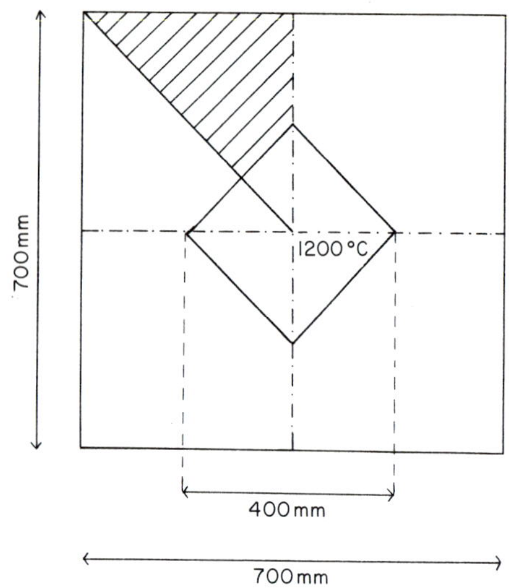

FIG. 8.1. The furnace wall.

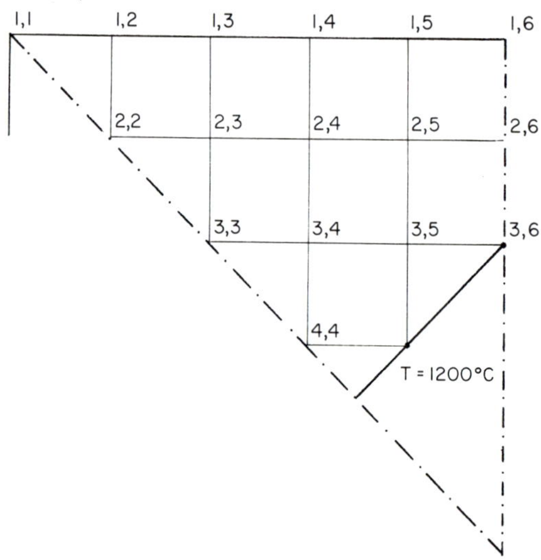

FIG. 8.2. The grid system.

depth is being ignored so that the problem is essentially 2-D steady-state with convective boundary conditions. Other details are:

Ambient temperature	$= 20\,^\circ$C
Surface coefficient	$= 10\,\text{W/m}^2\,\text{K}$
Thermal conductivity of furnace brick	$= 0{\cdot}7\,\text{W/mK}$

8.1.2. Type and Size of Grid

Since the furnace is symmetrical, it is only necessary to analyse a small segment (shown shaded in the diagram) and then multiply the segment heat loss by a factor of 8 to obtain the loss for the whole structure. The shape of the segment suggests that a square grid be employed for most of the area with a triangular grid for the area adjacent to the furnace inner wall. The resulting mesh, based on the coarse mesh size of 70 mm square, appears in Fig. 8.2.

8.1.3. Development of Finite Difference Equations

There is no need to develop FDEs for every node since all the internal nodes are described by the same equation and all the boundary nodes except one are also described by one type of equation; the exception is the corner node (T_{11}) which requires a special FDE. These three basic equations will now be derived using the energy balance concept.

(i) *Internal nodes*. Applying the energy balance to node (2,4) for example, as shown in Fig. 8.3, and remembering that the problem is 2-D steady-state,

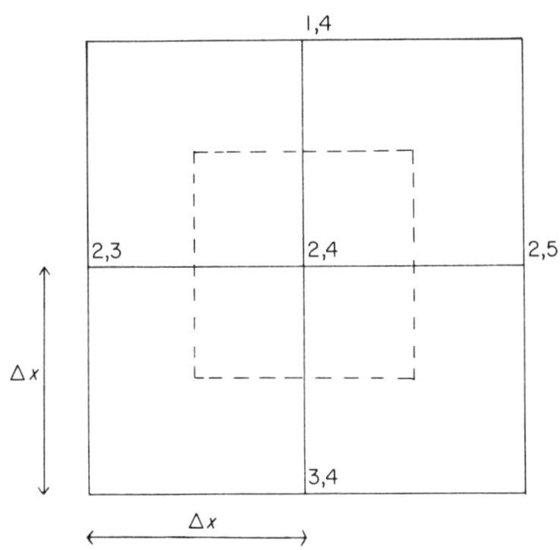

FIG. 8.3. Typical internal node.

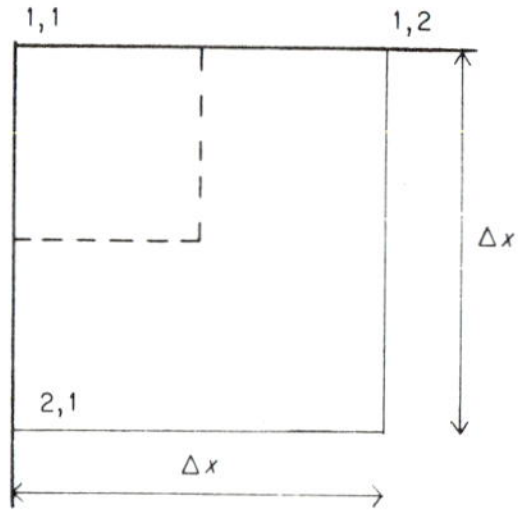

FIG. 8.4. Corner node.

with no internal heat generation and with constant thermal conductivity k, the net heat flows to node (2,4) from adjacent nodes will equal zero. Thus

$$\sum kA\frac{\Delta T}{\Delta x} = 0$$

that is,

$$\frac{kA}{\Delta x}[(T_{14} - T_{24}) + (T_{25} - T_{24}) + (T_{34} - T_{24}) + (T_{23} - T_{24})] = 0 \quad (8.1.1)$$

or

$$(T_{14} + T_{25} + T_{34} + T_{23} - 4T_{24}) = 0$$

(ii) *Corner node*. Energy flows to the corner node (1,1) by conduction and convection, the net sum of these flows being zero (see Fig. 8.4). Thus

$$hA\Delta T + \sum kA\frac{\Delta T}{\Delta x} = 0$$

or

$$h\Delta x(T_A - T_{11}) + \frac{k\Delta x}{2\Delta x}[(T_{21} - T_{11}) + (T_{12} - T_{11})] = 0$$

Using $h = 10\ \text{W/m}^2\text{K}$, $\Delta x = 0{\cdot}07\ \text{m}$, $k = 0{\cdot}7\ \text{W/mK}$, $T_{12} = T_{21}$ by symmetry, and $T_A = 20\,^\circ\text{C}$, the FDE for the corner node is therefore

$$(T_{12} - 2T_{11}) = -20 \quad (8.1.2)$$

(iii) *Other boundary nodes*. All the other boundary nodes away from the corner can be represented by the same type of FDE. Again, by summing the heat flows by convection and conduction to zero, we have, for node (1,3) of Fig. 8.5 for example:

$$h\Delta x(T_A - T_{13}) + \frac{k\Delta x}{\Delta x}(T_{23} - T_{13}) + \frac{k\Delta x}{2\Delta x}[(T_{12} - T_{13}) + (T_{14} - T_{13})] = 0$$

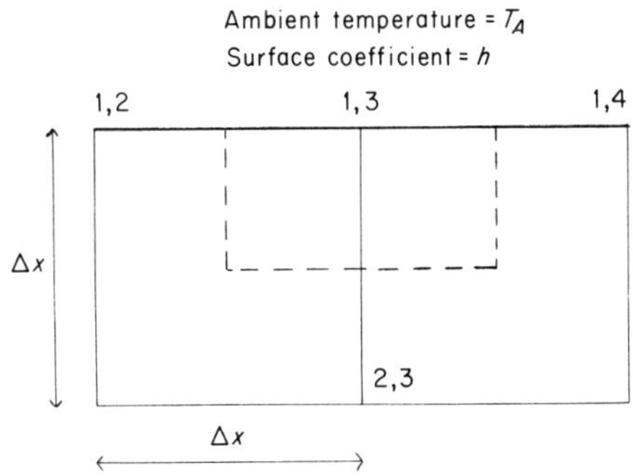

FIG. 8.5. Other boundary nodes.

Inserting the various numerical values as for the corner node gives the FDE

$$\tfrac{1}{2}(T_{12} + T_{14}) + T_{23} - 3T_{13} = -20 \qquad (8.1.3)$$

The FDEs for all the nodes are most easily represented in matrix form:

Node	T_{11}	T_{12}	T_{13}	T_{14}	T_{15}	T_{16}	T_{22}	T_{23}	T_{24}	T_{25}	T_{26}	T_{33}	T_{34}	T_{35}	T_{44}			Equation type
1,1	−2	+1														=	−20	(8.1.2)
1,2	$+\frac{1}{2}$	−3	$+\frac{1}{2}$				+1									=	−20	(8.1.3)
1,3		$+\frac{1}{2}$	−3	$+\frac{1}{2}$				+1								=	−20	,,
1,4			$+\frac{1}{2}$	−3	$+\frac{1}{2}$				+1							=	−20	,,
1,5				$+\frac{1}{2}$	−3	$+\frac{1}{2}$				+1						=	−20	,,
1,6					+1	−3					+1					=	−20	,,
2,2		+2					−4	+2								=	0	(8.1.1)
2,3			+1				+1	−4	+1			+1				=	0	,,
2,4				+1				+1	−4	+1			+1			=	0	,,
2,5					+1				+1	−4	+1			+1		=	0	,,
2,6						+1				+2	−4					=	−1200	,,
3,3								+2				−4	+2			=	0	,,
3,4									+1			+1	−4	+1	+1	=	0	,,
3,5										+1			+1	−4		=	−2400	,,
4,4													+2		−4	=	−2400	,,

(8.1.4)

8.1.4. Choice of Solution Procedure

The matrix equation set eqn. (8.1.4), being sparse, of reasonable size and with a dominant diagonal, may be solved by direct or iterative methods. Here we use the Gauss elimination technique as discussed in Section 7.2 with Fortran listing Program 2.1 of Chapter 7.

8.1.5. Computed Results and Discussion

The computer program yielded the results as shown schematically in Fig. 8.6.

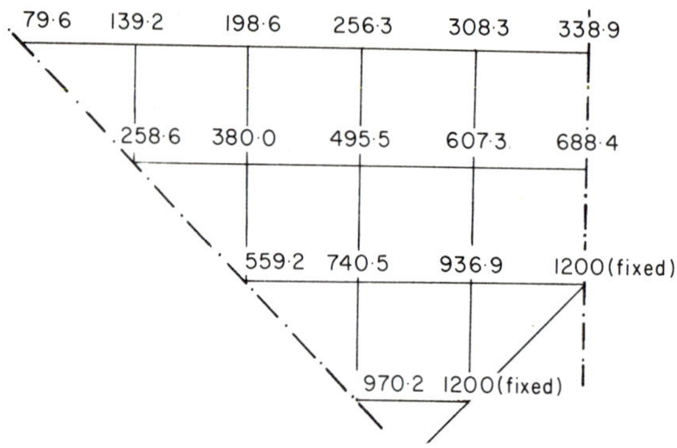

FIG. 8.6. Computed results.

The actual heat loss from the furnace could be calculated either by finding the average surface temperature and applying the simple convection equation or by using the conducting-rod analogy at the inner or outer boundaries:

(i) The average surface temperature

$$= [(\tfrac{1}{2}(79{\cdot}6 + 139{\cdot}2) + \tfrac{1}{2}(139{\cdot}2 + 198{\cdot}6) + \tfrac{1}{2}(198{\cdot}6 + 256{\cdot}3) + \tfrac{1}{2}(256{\cdot}3 + 308{\cdot}3) + \tfrac{1}{2}(308{\cdot}3 + 338{\cdot}9))/5]$$
$$= 222{\cdot}3\,^\circ\text{C}$$

Thus the heat flow from the outer surface by convection $= hA\Delta T$

$$= 10 \times 2{\cdot}8 \times (222{\cdot}3 - 20)$$
$$= 5{\cdot}67\,\text{kW/m height of furnace}$$

(ii) Using the conducting-rod analogy, and remembering that some rods only represent half of one square due to the subdivision of the object, the heat flow into the furnace segment

$$= \frac{k}{2}(1200 - 688{\cdot}4) + k(1200 - 936{\cdot}9) + k(1200 - 936{\cdot}9) + k(1200 - 970{\cdot}2)$$
$$= 1011{\cdot}8k \text{ W/segment/m height of furnace}$$
$$= 1011{\cdot}8 \times 0{\cdot}7 \times 8 \text{ for the whole cross-section}$$
$$= 5{\cdot}67\,\text{kW/m height of furnace}$$

8.2. A BUS-BAR PROBLEM: 2-D STEADY-STATE HEAT FLOW WITH INTERNAL HEAT GENERATION

8.2.1. Problem Statement

The section of the bus-bar is shown in Fig. 8.7, with the dimension in the third direction equal to 2 m. Under overload conditions the bus-bar carries a current of 54 000 A. The problem is to derive the steady-state temperature

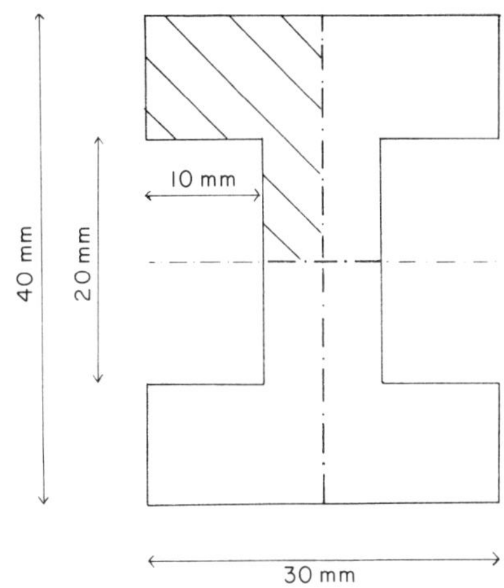

FIG. 8.7. The bus-bar.

distribution within the bar when the outside is maintained at 50 °C. Since the third dimension is so much greater than the other two, the problem may be considered 2-D steady-state with uniform internal heat generation and known temperatures at the boundary. Values of specific constants for the bus-bar material (copper) are

Electrical resistivity $= 2 \times 10^{-8}\ \Omega/\text{m}$
Thermal conductivity $= 381$ W/mK

8.2.2. Type and Size of Grid

Since the section is symmetrical, it is only necessary to analyse a segment of the bar (shown shaded in Fig. 8.7) and multiply the segment heat loss by a factor of 4 to obtain the heat loss for the whole structure. The shape of the segment suggests that a square grid be employed throughout. The resulting mesh, based on the most coarse possible mesh size of 5 mm square, appears in Fig. 8.8.

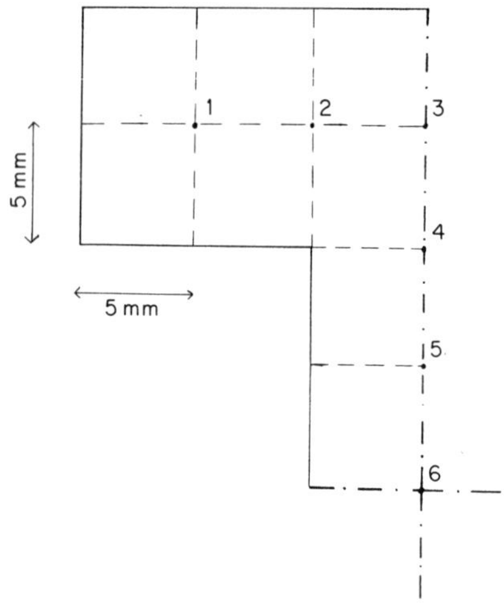

FIG. 8.8. The grid system.

8.2.3. Development of Finite Difference Equations

Since the temperatures at the boundary are specified, only FDEs for the internal nodes will be required and all these will in fact be of the same form. The basic equation will now be derived using the energy balance concept.

Applying the energy balance to node 2, for example, as shown in Fig. 8.9, and remembering that the problem is 2-D steady-state with constant thermal conductivity and internal heat generation, we know that the heat flow to node 2 from adjacent nodes added to the heat generated in the control volume represented by node 2 will equal zero. Thus

$$\sum kA\frac{\Delta T}{\Delta x} + H(\Delta x)^2 = 0$$

that is,

$$\frac{kA}{\Delta x}[(50 - T_2) + (T_1 - T_2) + (T_3 - T_2) + (50 - T_2)] + H(\Delta x)^2 = 0$$

Since we have heat generated/unit volume = resistivity × (current density)^2

$$H = 2 \times 10^{-8} \times (54\,000/(8 \times 10^{-4}))^2$$
$$= 9{\cdot}112 \times 10^7 \text{ W/m}^3$$

Further substituting the values for $k = 381$ W/mK, $A = \Delta x \cdot 1$, $\Delta x = 0{\cdot}05$m, will give the FDE:

$$T_1 + T_3 - 4T_2 + 106 = 0$$

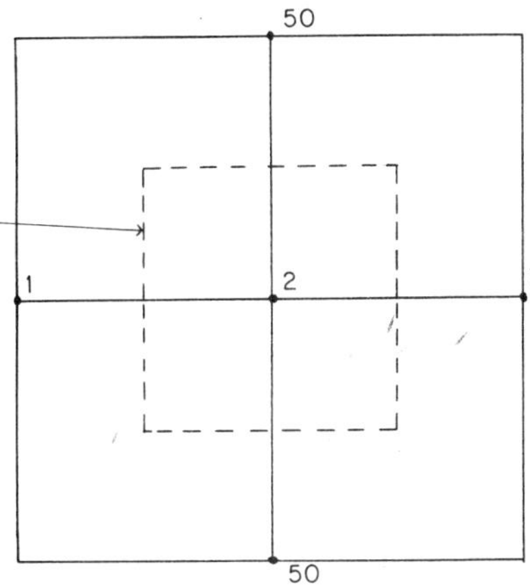

FIG. 8.9. Typical internal node.

In matrix form the FDE set is:

Node	T_1	T_2	T_3	T_4	T_5	T_6	
1	−4	1					= −156
2	1	−4	1				= −106
3		2	−4	1			= − 56
4			1	−4	1		= −106
5				1	−4	1	= −106
6					2	−4	= −106

8.2.4. Choice of Solution Procedure

The above matrix shows the tridiagonal nature of the equation set, and although the equation set is small in number it is good practice to choose the most appropriate solution technique. A direct method, such as Gauss elimination, applied to a tridiagonal equation set is the best and the appropriate computer program using the resulting TDMA is discussed in detail in Section 7.2, along with Program 2.2, of Chapter 7.

8.2.5. Computed Results and Discussion

The program yielded the following results:

$$T_1 = 52{\cdot}25\,°C$$
$$T_2 = 53{\cdot}02\,°C$$
$$T_3 = 53{\cdot}81\,°C$$
$$T_4 = 53{\cdot}22\,°C$$
$$T_5 = 53{\cdot}06\,°C$$
$$T_6 = 53{\cdot}03\,°C$$

The actual heat loss from the bus-bar is known in this case since in the steady-state the heat loss must balance the internal heat generation. Considering a 1 m length of bar, the volume of this metal is

$$1 \times (0{\cdot}03 \times 0{\cdot}04 - 0{\cdot}02 \times 0{\cdot}02) = 0{\cdot}0008\,\text{m}^3$$

The actual heat generated within the bar, assuming uniform heating,

$$= (0{\cdot}0008 \times H)\,\text{W/m}$$

where $H = 9{\cdot}112 \times 10^7\,\text{W/m}^3$, and so this reduces to

$$72{\cdot}9\,\text{kW/m}$$

Thus the heat loss from the bar, if calculated by some other method, should equal 72·9 kW.

Applying the conducting-rod analogy over the whole section suggests a heat flow in terms of conductivity k of $(89{\cdot}96\,k)$ W/m = 34·3 kW/m. This massive error illustrates two important points:

(i) The initial choice of grid size is usually based on the idea of 'the coarsest that can be got away with' to get a feel for the solution. It is often a sound idea subsequently to reduce the grid size and compare the two sets of solutions, possibly using extrapolation as described in Chapter 4, Section 4.4.

(ii) In problems involving internal heat generation the temperature gradients near the boundary are usually very much greater than elsewhere in the body, particularly when the thermal diffusivity is high. This requires a fine mesh to be employed so as to have sufficient node points near the boundary. Additional error is involved because of the use of one-sided derivatives near the boundary.

Bearing these two points in mind, the analysis is now repeated using a grid of half the previous size, that is, $\Delta x = 2{\cdot}5$ mm. The computed results using the Gauss elimination solution scheme described in Section 7.2 and Program 2.1 of Chapter 7 are shown schematically for the upper right-hand quadrant in Fig. 8.10.

Using the conducting-rod analogy, the segment heat flow is

$$(32{\cdot}585k)\,\text{W/m length}$$

where k is the conductivity. Thus the heat flow from the whole section is

$$(32{\cdot}585k) \times 4\,\text{W/m} = 49{\cdot}65\,\text{kW/m}$$

The finer grid obviously produces a much more accurate solution. In the

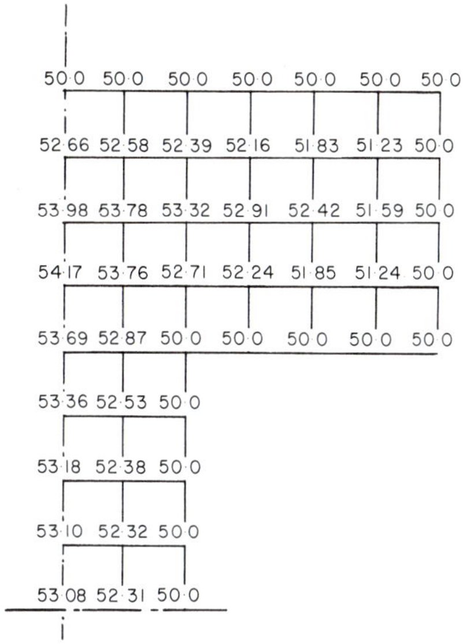

FIG. 8.10. Computed results with fine mesh.

problem as stated, the temperature at the boundary was actually specified. Normally the boundary values would have to be derived like any other temperature in the section. The effect of specifying the boundary temperatures is to reduce the actual area of the section represented by the nodes, and there is a 'layer' (of thickness equal to half the mesh size) around the entire outer boundary not represented in the FDE derivations. The effect of this in turn is to reduce the quantity of internal heat generation represented by the FDEs and so the predicted temperatures are therefore slightly too low.

8.3. A BUS-BAR PROBLEM: SURGE CURRENT CONDITIONS, 2-D TRANSIENT HEAT FLOW WITH INTERNAL HEAT GENERATION

8.3.1. Problem Statement

Occasionally, bus-bars are subjected to large surge currents for short periods; in this case 20 000 A for 1 s. British Standard 159 (1957) for bus-bars and bus-bar connections requires that the surface temperature of the bus-bar shall not exceed ambient temperature by more than 150 °C under these conditions. The objective of this analysis is to determine the actual rise in temperature of the surface, and the extent to which this rise can be reduced by more vigorous cooling at the boundary.

The bus-bar considered here is rectangular and measures 25·4 mm by 6.35 mm and is 2 m long; the problem is essentially 2-D. Under steady-state conditions the current is 250 A. The bus-bar material is copper having the following properties:

Thermal conductivity	= 381 W/mK
Electrical resistivity	$= 2 \times 10^{-8}\,\Omega/\text{m}$
Specific heat	= 383 J/kg K
Density	$= 8954\ \text{kg/m}^3$

8.3.2. Type and Size of Grid

The shape of the cross-section suggests a square mesh. As the section is symmetrical about two axes, only a quarter of the section need be analysed as shown in Fig. 8.11. In view of the findings of the analogous problem in

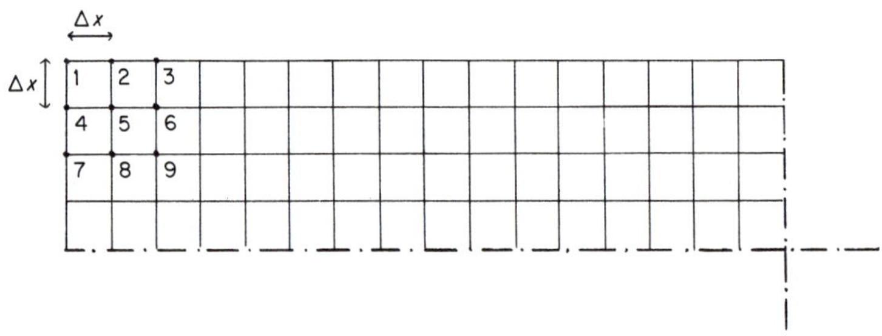

FIG. 8.11. Square mesh covering quarter section of rectangular bus-bar.

Section 8.2 concerning the use of too coarse a grid, on this occasion a reasonably fine grid will be used straight away. A mesh size of 0·794 mm will generate the node formation as shown in Fig. 8.11, where only a few of the nodes are labelled.

8.3.3. Development of Finite Difference Equations

Under steady-state conditions it has already been seen in the problem of Section 8.2 that the temperature variation across the section is very small, so that the whole section will be taken to operate at a temperature difference ΔT above ambient.

If the heat is transferred from the bus-bar by turbulent free convection, then the heat transfer coefficient is given by (see Eckert & Drake, 1959)

$$h = 1{\cdot}31(\Delta T)^{1/3}$$

Under steady-state conditions there will be a balance between the rate at which heat is generated internally and the rate at which it is convected away:

$$\text{Rate of internal generation} = \text{Resistivity} \times (\text{Current density})^2 \times \text{Volume of bus-bar} = 15{\cdot}50\ \text{W}$$

$$\text{Rate of heat transfer by convection} = hA\Delta T$$

where A, the surface area of the bar, $= 0{\cdot}127\ \text{m}^2$. Substituting for h in the above equation yields a value for ΔT equal to $29{\cdot}99\ °\text{C}$. Repeating the calculation for forced convective cooling yields a value of $0{\cdot}72\ °\text{C}$ for ΔT.

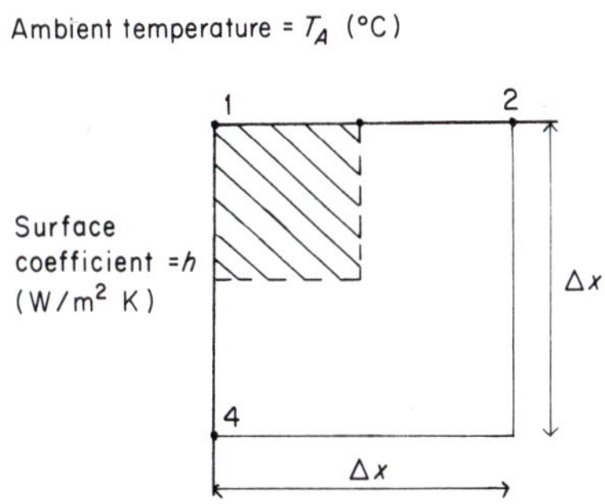

FIG. 8.12. Corner node.

There are three typical groups of node points which will require different types of finite difference equation: the corner nodes (*e.g.* node 1 of Fig. 8.12), the boundary nodes (*e.g.* node 2 of Fig. 8.13) and the internal node (*e.g.* node 5 of Fig. 8.14).

(i) *FDE for a corner node* (*node* 1). Applying the energy balance concept to the control volume for node 1 shown in Fig. 8.12:

$$\begin{matrix}\text{Net heat flow to node 1 by} \\ \text{conduction and convection}\end{matrix} = \begin{matrix}\text{`Mass' of node} \times \text{Specific heat} \\ \times\ \text{Rate of temperature change}\end{matrix}$$

Thus

$$k\left(\frac{\Delta x}{2}\right)\left(\frac{T_2 - T_1}{\Delta x}\right) + k\left(\frac{\Delta x}{2}\right)\left(\frac{T_4 - T_1}{\Delta x}\right) + \frac{H(\Delta x)^2}{4} + h\Delta x(T_A - T_1) = \rho C_p \frac{(\Delta x)^2}{4}\frac{T_1' - T_1}{\Delta t}$$

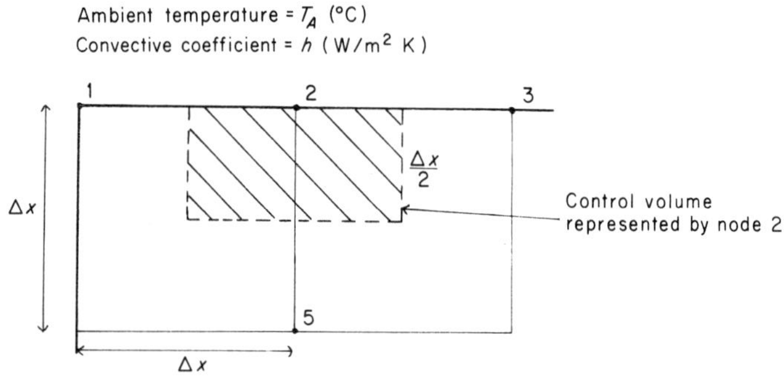

FIG. 8.13. Typical boundary node.

Rearrangement gives the new temperature T_1' at node 1 as

$$T_1' = \text{Fo}\left[2T_2 + 2T_4 + 4\text{Bi}\cdot T_A + \frac{H(\Delta x)^2}{k} + T_1\left(\frac{1}{\text{Fo}} - 4\,\text{Bi} - 4\right)\right]$$

where the Fourier number $\text{Fo} = \dfrac{k}{\rho C_p}\cdot\dfrac{\Delta t}{(\Delta x)^2}$

(ii) *FDE for a boundary node* (*node* 2). Applying the energy balance concept to node 2 as shown in Fig. 8.13:

$$\left[k\left(\frac{\Delta x}{2}\right)\frac{T_1 - T_2}{\Delta x} + k\left(\frac{\Delta x}{2}\right)\frac{T_3 - T_2}{\Delta x} + H\frac{(\Delta x)^2}{2} + k(\Delta x)\frac{(T_5 - T_2)}{\Delta x} + h(\Delta x)(T_A - T_2)\right] = \rho C_p\frac{(\Delta x)^2}{2}\frac{T_2' - T_2}{\Delta t}$$

giving the new temperature T_2' at node 2 as

$$T_2' = \text{Fo}\left[2T_5 + T_1 + T_3 + T_2\left(\frac{1}{\text{Fo}} - 4 - 2\,\text{Bi}\right) + 2T_A\text{Bi} + H\frac{(\Delta x)^2}{k}\right]$$

(iii) *FDE for an internal node* (*node* 5). Applying the energy balance to node 5 as shown in Fig. 8.14 will generate the following FDE for the new temperature at node 5 after a time-interval Δt:

$$T_5' = \text{Fo}\left[T_2 + T_4 + T_6 + T_8 + T_5\left(\frac{1}{\text{Fo}} - 4\right) + H\frac{(\Delta x)^2}{k}\right]$$

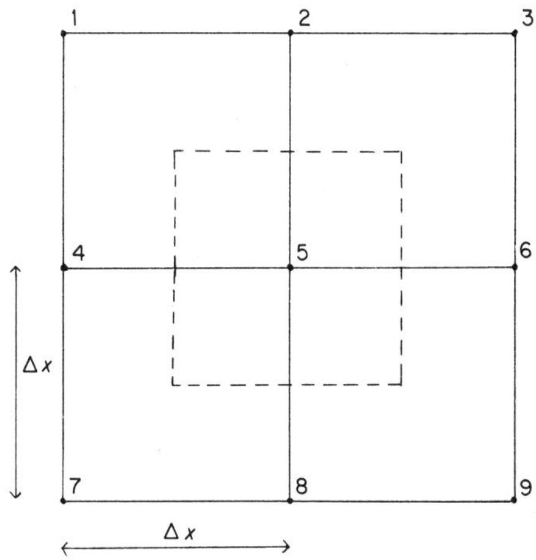

FIG. 8.14. Typical internal node.

where Fo is the Fourier number as defined above. For stability, it has been established that negative coefficients in the FDE should be avoided. The most stringent criterion (*i.e.* the minimum value for Fo) is generated by the FDE for the corner node (node 1). From this equation,

$$\text{Fo} \leq \frac{1}{(4 + 4\,\text{Bi})}$$

where the Biot number Bi is

$$\text{Bi} = \frac{h\Delta x}{k}$$

$$= \frac{6{\cdot}97\,(0{\cdot}794 \times 10^{-3})}{381} = 1{\cdot}45 \times 10^{-5}$$

where the heat transfer coefficient h is derived from the turbulent free convection relationship based on the maximum temperature difference of 150 °C. The resulting critical Fourier number is equal to 0·25, which permits evaluation of the critical time interval Δt as

$$\Delta t \leq \frac{\text{Fo}\,(\Delta x)^2}{(k/\rho C_p)}$$

that is, Δt should not exceed a value of $1{\cdot}419 \times 10^{-3}$ s. As the period of the surge is 1 s, the number of cycles of calculation required is

$$\frac{1}{\Delta t} = 704$$

8.3.4. Choice of Solution Procedure

Since the equations have been derived explicitly, that is, in the form that one unknown temperature is related to several previous time-level known temperatures, the solution procedure itself will be explicit in nature. This type of solution procedure is fully explained in Chapter 6 and Section 7.5 of Chapter 7, with an illustrative program, Program 5.1 of Chapter 7.

8.3.5. Computed Results and Discussion

The computed results are shown in Fig. 8.15. The hottest point on the bus-bar surface lies at the middle of the longest side of the section. It will be

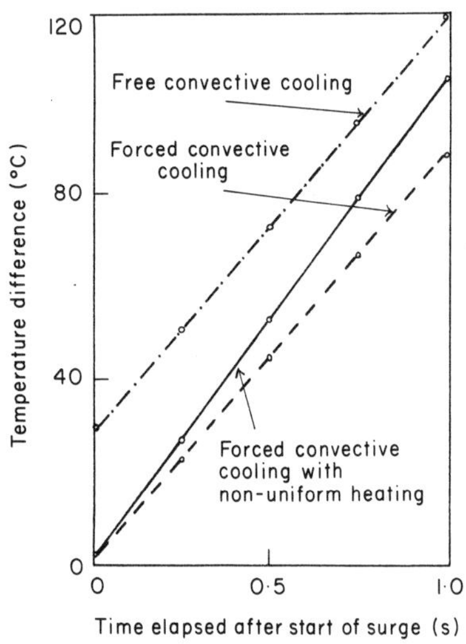

FIG. 8.15. Temperature history of surface 'hot-spot' for different boundary conditions.

seen from the figure that the advantage in forced convective cooling is that the temperature differential established in the steady state is maintained during the surge and the bar would therefore be allowed to withstand the surge current for about 1·5 s. The cooling mode clearly has no effect during the transient condition since the rate of internal heat generation is of an order several times greater than the rate of cooling.

It is believed that under these conditions of high current density, the flow of current is not uniform over the whole cross-section. At low frequencies the current is said to flow through a core inside the cross-section of the bar away from the outer boundary. The computer model can easily be adapted to describe this phenomenon, and as an illustration the program was run with no heat generation term allocated to boundary nodes. Since the same current is flowing through a smaller cross-sectional area, the current density and hence the rate of energy generation will increase. The predicted temperature response is shown in Fig. 8.15. The increased rate of energy generation is clearly reflected by the more rapid rate of temperature change.

8.4. TRANSIENT HEAT FLOW THROUGH A CIRCULAR HEATING ELEMENT

8.4.1. Problem Statement

Figure 8.16 shows the cross-section of a circular heating element; the heating coil is surrounded by electrical insulation material and the whole is

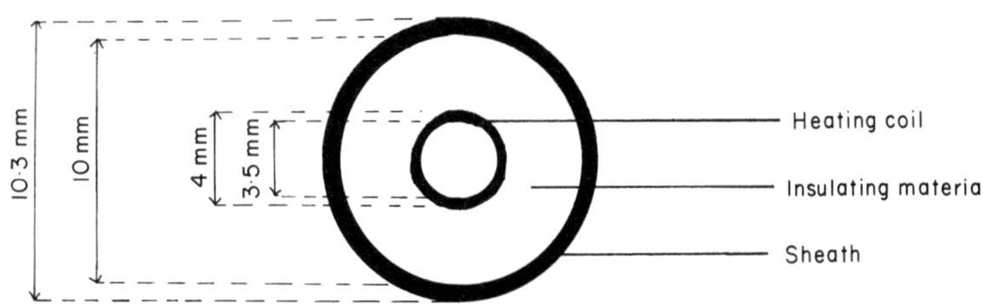

FIG. 8.16. Cross-section of circular heating element.

encased in a thin metal sheath. This particular element is to be used for heating water. Thermal properties of the insulating material are taken as:

Specific heat = 1300 J/kg K
Density = 3050–3150 kg/m^3
Thermal conductivity = 1·8–7·0 W/mK

The insulating material is in the form of a compacted powder, and herein lies the problem, since the thermal conductivity is greatly affected by the degree of compaction, varying from a value of 1·8 to 7·0 W/mK. The transient behaviour of the element is therefore of particular interest, since for domestic applications it is not acceptable to have too great a heating-up time. The problem to be solved is therefore one of axisymmetric heat flow with internal heat generation. From the dimensions shown in the diagram, it is clear that the length of the heating element is very much greater than the diameter and therefore temperature variations along the element will be ignored.

The convective heat transfer conditions will be assumed to follow the characteristic boiling curve. Using the data available from Eckert & Drake (1959), the heat transfer coefficients which apply at the boundary will be taken as:

(i) For the difference $(T_B - T_A) \leq 10\,^\circ\mathrm{C}$,

$$h = 5000\ \mathrm{W/m^2K}$$

(ii) For the difference $(T_B - T_A) > 10\,^\circ\mathrm{C}$,

$$h = 1/3[500(T_B - T_A) + 10^4]\ \mathrm{W/m^2K}$$

8.4.2. Type and Size of Grid

Clearly a grid comprising circular elements is required. Since the problem is axisymmetric, the grid need only comprise a set of concentric circles. The simplest grid will be one in which the difference between the diameters of adjacent circles is a constant. The size and position of the heating coil suggests that the most appropriate size of grid (for the initial calculations)

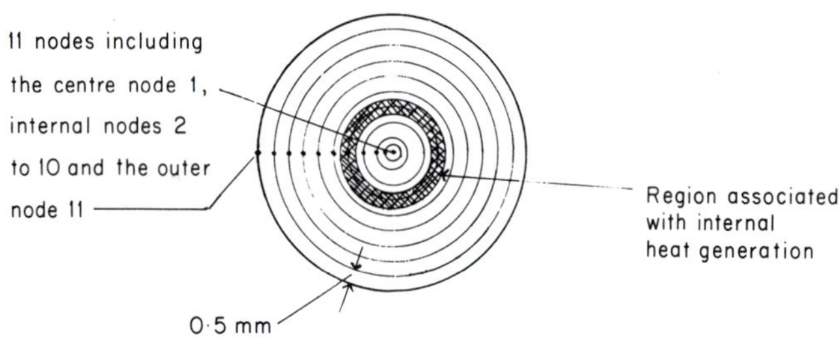

FIG. 8.17. Grid system.

would be one in which the two adjacent nodes just encompassed the heating coil, as shown in Fig. 8.17.

The sheath material is not included in the figure for two reasons:

(i) The thermal diffusivity of the sheath material is approximately 15 times greater than that of the insulating material.

(ii) The sheath is very thin compared with the other dimensions of the heating element, and therefore the total thermal inertia of the sheath is not a significant factor in the transient behaviour of the element.

8.4.3. Development of the Finite Difference Equations

The problem requires three different types of FDE to be derived: one for the boundary node, one for the centre-line node and a general one for the

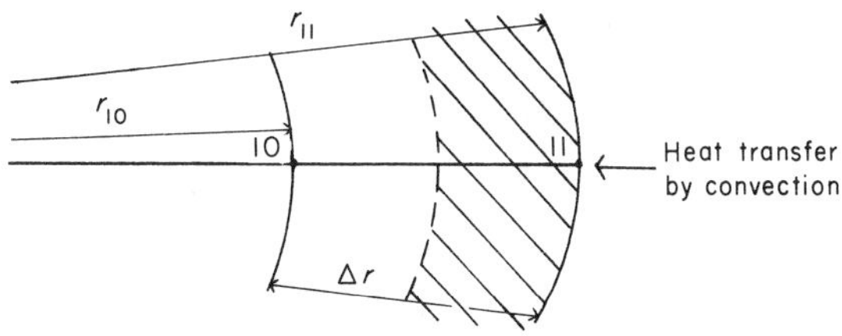

FIG. 8.18. Boundary node.

remainder of the internal nodes. In addition, the internal heat generation term H has to be evaluated.

(i) *FDE for boundary node* (*node* 11). Applying the energy balance concept to node 11 as shown in Fig. 8.18:

(Net heat flow to node 11 by conduction and convection)
+ (Internal heat generation)
= 'Mass' of element × Specific heat × Rate of temperature change

i.e.

$$\left[k\left(\frac{T_{10}-T_{11}}{\Delta r}\right)\cdot 2\pi\left(\frac{r_{10}+r_{11}}{2}\right)+h(2\pi r_{11})(T_A-T_{11})\right]$$
$$=\rho\pi(r_{11}^2-r_{10}^2)C_p\left(\frac{T_{11}'-T_{11}}{\Delta t}\right)$$

Since $r_{11}^2 - r_{10}^2 \approx 2\,r_{11}\,\Delta r$ this equation resolves to

$$T_{11}'=\frac{\text{Fo}}{2}\left\{\left(2-\frac{\Delta r}{r_{11}}\right)T_{10}+\frac{2h\,\Delta r}{k}\,T_A+T_{11}\left[\frac{2}{\text{Fo}}-\left(2-\frac{\Delta r}{r_{11}}+\frac{2\,h\,\Delta r}{k}\right)\right]\right\}$$

where the Fourier number $\text{Fo}=\dfrac{k}{\rho C_p}\cdot\dfrac{\Delta t}{(\Delta r)^2}$.

(ii) *FDE for centre-line node* (*node* 1 *of Fig.* 8.17). The area which is represented by node 1 is 0·25 % of the whole cross-sectional area of the heating element. The true criterion to apply at the centre-line is that the temperature gradient $\partial T/\partial r$ is zero. The simplest but not the most accurate way to implement this constraint is to make the temperature gradient between node 1 and the adjacent node equal to zero. In finite difference terms this means that any new temperature at node 2 will be calculated explicitly from temperatures at nodes 1, 2 and 3 and the new temperature at node 1 then made equal to it.

(iii) *FDE for the remainder of internal nodes* (*nodes* 2 *to* 10 *inclusive of Fig.* 8.17). As an example, the FDE will be derived directly from the general differential equation for constant thermal conductivity:

$$k\left(\frac{\partial^2 T}{\partial r^2} + \frac{1}{r}\frac{\partial T}{\partial r} + \frac{1}{r^2}\frac{\partial^2 T}{\partial \theta^2} + \frac{\partial^2 T}{\partial z^2}\right) + H = \rho C_p \frac{\partial T}{\partial t}$$

which resolves to

$$k\left(\frac{\partial^2 T}{\partial r^2} + \frac{1}{r}\frac{\partial T}{\partial r}\right) + H = \rho C_p \frac{\partial T}{\partial t}$$

for the axisymmetric heat flow of a system whose length is considerably greater than its diameter with uniform heating. Replacing the above equation with its finite difference equivalent and applying it to node 5, say:

$$k\left[\frac{T_4 - 2T_5 + T_6}{(\Delta r)^2} + \frac{1}{r_5}\left(\frac{T_4 - T_6}{2\,\Delta r}\right)\right] + H_5 = \rho C_p\left(\frac{T_5' - T_5}{\Delta t}\right)$$

giving the new temperature T_5' after a time-interval Δt as

$$T_5' = \text{Fo}\left[T_4\left(1 + \frac{\Delta r}{2r}\right) + T_6\left(1 - \frac{\Delta r}{2r}\right) + T_5\left(\frac{1}{\text{Fo}} - 2\right) + H_5\frac{(\Delta r)^2}{k}\right]$$

where the Fourier number $\text{Fo} = \dfrac{k}{\rho C_p}\cdot\dfrac{\Delta t}{(\Delta r)^2}$.

There is one important difference between the equation for node 5 and those for all the other internal nodes: the internal heat generation from the coil will occur only at the area represented by node 5, and therefore this part of the equation (*i.e.* $H(\Delta r)^2$) will be set to zero for all the other nodes.

(iv) *Evaluation of internal heat generation term H*. Heat elements such as the one in this problem are normally rated in terms of power per unit surface area, *i.e.* W/m². For this problem the rating is $3{\cdot}27 \times 10^5$ W/m² surface area. For an element of length 1·0 m and diameter 10 mm

$$\begin{aligned}\text{Actual energy released} &= 3{\cdot}27 \times 10^5 \times (1 \times \pi \times 10/1000)\\ &= 10{\cdot}28 \times 10^3\,\text{W}\end{aligned}$$

This energy is released in the volume associated with the heating element, *i.e.* the control volume associated with node 5, which is equal to

$$\pi(2{\cdot}5^2 - 2{\cdot}0^2) \times 10^{-6} \times 1$$

i.e.

$$V = 7{\cdot}069 \times 10^{-6}\,\text{m}^3$$

$$\therefore H = \frac{\text{Power}}{\text{Unit volume}} = \frac{10{\cdot}28 \times 10^3}{7{\cdot}069 \times 10^{-6}} = 1{\cdot}454 \times 10^9\,\text{W/m}^3$$

This value of H appears as H_5 in the FDE for node 5.

The equations have been derived explicitly and are therefore subject to the stability criterion of avoiding negative coefficients (see Chapter 4, Section 4.5). The most critical stability criterion occurs in the FDE for the boundary node, which requires the Fourier number Fo to be restricted by

$$\mathrm{Fo} \leq 1 \bigg/ \left(2 - \frac{\Delta r}{r_{11}} + \frac{2h\,\Delta r}{k}\right)$$

Assuming a maximum heat transfer coefficient of 20 000 W/m^{2}K, the Biot number Bi

$$= \frac{h\,\Delta r}{k} = \frac{20\,000{\cdot}0 \times 0{\cdot}0005}{1{\cdot}8} = 5{\cdot}55$$

$$\frac{\Delta r}{r_{11}} = \frac{0{\cdot}0005}{0{\cdot}005} = 0{\cdot}10$$

$$\therefore\ \mathrm{Fo} \leq \frac{1}{13{\cdot}20}$$

The time-step Δt is given by

$$\Delta t = \mathrm{Fo}\frac{(\Delta r)^2}{k}\rho C_{\mathrm{p}}$$

Using the values of ρ and k for the 'poorly compacted' condition will give a value for Δt of 0·0417 s.

8.4.4. Choice of Solution Procedure

Since the equations have been derived explicitly, that is, in the form that one unknown temperature is related to several previous time-level known temperatures, the solution procedure itself will be explicit in nature. This type of solution procedure is fully explained in Chapter 6 and Section 7.5 of Chapter 7, with an illustrative program, Program 5.1 of Chapter 7.

8.4.5. Computed Results and Discussion

Figure 8.19 shows the transient response of the heating element when the insulating material is 'well' and 'poorly' compacted. The advantages of good compaction are clear to see. The rather strange-looking kink in the 'poor' compaction curve indicates the effect of the lower-conductivity material in reducing the rate at which the internal heat generation is allowed to diffuse to the surface.

Another pronounced effect of poor compaction is that it greatly increases the temperatures to which the interior of the heating element is subjected. Figure 8.20 shows the temperature distribution through the element (for

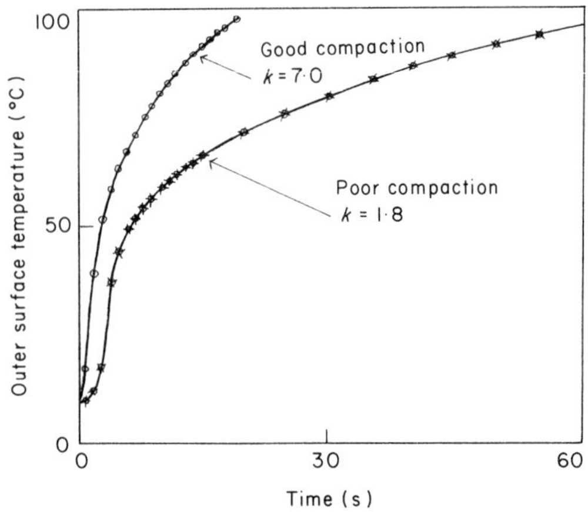

FIG. 8.19. Temperature of outer surface during 'start-up' for different degrees of compaction.

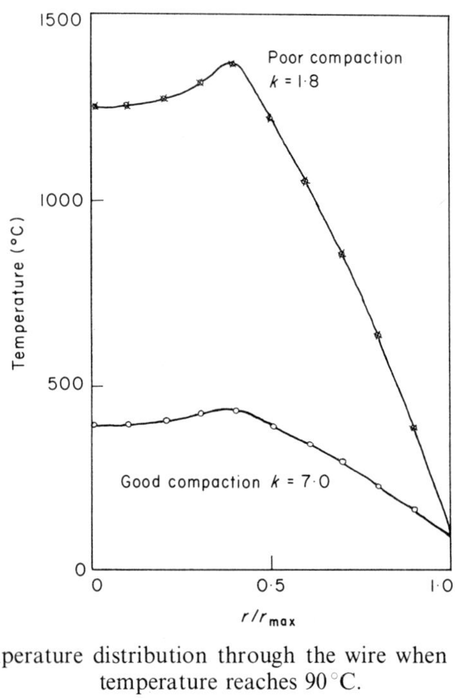

FIG. 8.20. Temperature distribution through the wire when the outer surface temperature reaches 90 °C.

both degrees of compaction) at the point at which the outside temperature has reached 90 °C. The choice of 90 °C as the common outer surface temperature is not especially significant, except that as the surface temperature rises, so will the bulk temperature of the water. In this calculation the ambient temperature has been assumed constant at 10 °C. If the heating element actually reached 100 °C with the ambient at 10 °C, film boiling would occur and the convective heat transfer coefficient would decrease rapidly. This would have the effect of suddenly raising the outer surface temperature—a contingency not allowed for in the computer model. The choice of outer surface temperature then is arbitrarily set at 90 °C—the value chosen serves to make the following point anyway: since the time taken to reach 90 °C is 50 s for the poorly compacted case and 15 s for the other case, the temperature distributions are widely different and clearly much more heat is actually stored within the poorly compacted element.

In conclusion, it can be seen quite clearly that the result of poor compaction is a reduction in response time, and perhaps worse, from the point of view of efficient use of energy, more energy is stored within the element—energy which will subsequently be lost when the heating element is switched off. It is a simple matter to evaluate the extent of this wastage. By summing the heat stored in each segment of the whole element and comparing the totals for the differently compacted systems, the difference in heat content is found to be approximately 0·15 MJ.

8.5. HEAT FLOW THROUGH A TRIANGULAR FIN: TRANSIENT AND STEADY-STATE BEHAVIOUR

8.5.1. Problem Statement

Figure 8.21 shows the section through a block of triangular fins. Initially the block is at a uniform temperature of 10 °C. The temperature of the inner

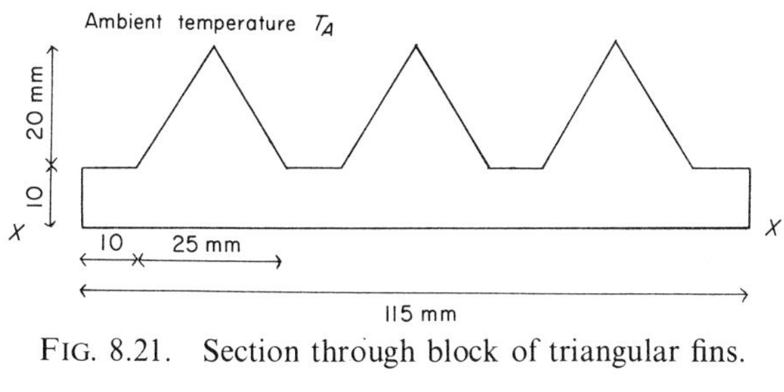

FIG. 8.21. Section through block of triangular fins.

surface (X-X on the figure) is suddenly raised to 1000 °C and the problem is to derive the temperature distribution in the fin after 30 s have elapsed and also the eventual steady-state distribution.

The fin material has the following properties:

Density	$= 4000\,kg/m^3$
Thermal conductivity	$= 200\,W/mK$
Specific heat	$= 2{\cdot}0\,kJ/kg\,K$

The upper surfaces of the fins are cooled by convection, a coefficient of $100\,W/m^2K$ applying. Since the aim of the problem is to illustrate the problems associated with triangular meshes, the dimension in the third direction will be ignored and the problem treated as two-dimensional.

8.5.2. Type and Size of Grid

The shape of the fin has been chosen to illustrate the application of triangular grids. Previous discussion on such grids in Section 4.2 of Chapter 4 showed that the method is only valid if no angle within a triangular element is greater than 90°. Working on the philosophy of choosing a coarse grid initially and then refining the grid later if necessary, the mesh shown in Fig. 8.22 is a reasonable first try.

8.5.3. Development of Finite Difference Equations

It would be possible to solve the problem by analysing the transient behaviour and then allowing the computer program to 'run on' until a

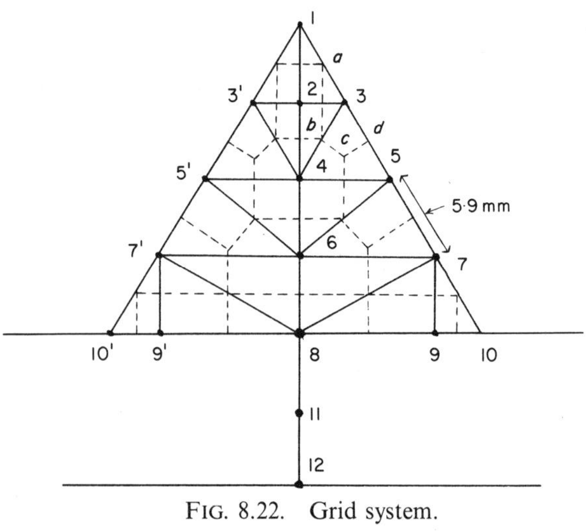

FIG. 8.22. Grid system.

TABLE 8.2

THE GEOMETRIC FACTORS AND CONTROL AREAS ASSOCIATED WITH EACH NODE

	1	2	3 3′	4	5 5′	6	7 7′	8
1		0·625	0					
2	0·625		1·6	0·625				
3 or 3′	0	1·6		0·3125	0·3125			
4		0·625	0·3125		0·65	1·25		
5 or 5′			0·3125	0·65		0·3125	0·625	
6				1·25	0·3125		0·325	1·875
7 or 7′					0·625	0·325		0

Additionally, $F_{7\,10} = 0$, $F_{97} = 1{\cdot}25$

Node	1	2	3	4	5	6	7
Control area (mm^2)	3·91	15·76	9·20	29·6	19·15	44·60	20·50

steady-state situation is reached. Apart from the waste of computer time, this approach will save time overall, particularly if the steady-state solution is not of prime interest. Here we shall solve the transient and steady-state problems independently. Energy balance equations using geometric factors (see Section 4.2 of Chapter 4) can be derived in the same general way as for square mesh systems, but it should be remembered that the control area represented by a node in a triangular grid will be an irregular polygon, such as area *abcd* for node 3.

Table 8.2 shows the calculated values for the geometric factors and control areas associated with each node.

(i) *Steady-state FDEs.* Applying the energy balance to node 1 for example, the net heat flow by conduction and convection to the node will be zero, *i.e.*

$$kF_{21}(T_2 - T_1) + kF_{31}(T_3 - T_1) + kF_{3'1}(T_{3'} - T_1) + hA(T_A - T_1) = 0$$

The geometric factors tabulated in Table 8.2 are now calculated in detail for node 1.

From the geometry of triangles 123 and 123′, the angles are as follows:

$$\text{angle } 123 = 90^\circ; \quad \text{angle } 132 = 58^\circ; \quad \text{angle } 13'2 = 58^\circ$$

$$\therefore F_{31} = \frac{1}{2\tan 90^\circ} = 0 \ (= F_{3'1} \text{ by symmetry})$$

In triangle 123 the geometric factor F_{21} will be given by

$$F_{21} = \frac{1}{2 \tan 58^\circ} = 0{\cdot}3125$$

A similar value for F_{21} will obviously evolve from triangle 123′, giving a total value for F_{21} from all sources of $(0{\cdot}3125 + 0{\cdot}3125) = 0{\cdot}625$.

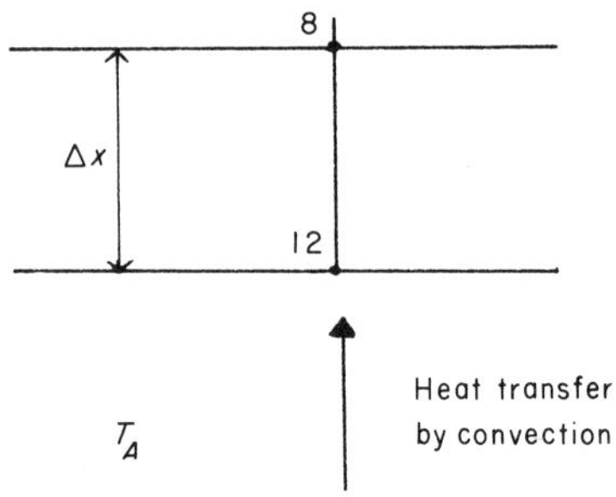

FIG. 8.23. 1-D heat flow from lower face of block to base of fin.

Inserting the numerical values of k, the various F's, and a value of 10 °C for the ambient temperature, into the FDE for node 1 gives

$$0{\cdot}995T_2 - T_1 = -0{\cdot}047$$

It will be assumed that the variation in temperature across the base of the fin is small and that the heat flow from the lower face of the block to the base of the fin is essentially one-dimensional; this assumption permits a simple one-dimensional heat balance of conduction and convection across the area below the fin, as illustrated in Fig. 8.23:

$$\frac{kA}{\Delta x}(T_8 - T_{12}) = hA(T_{12} - T_A)$$

giving $T_8 = 995$ °C. The temperatures at nodes 8, 9, and 10 will therefore be taken as 995 °C.

In matrix form the equation set appears as follows:

Node	T_1	T_2	T_3	T_4	T_5	T_6	T_7		
1	−1	0·9953						=	0·047
2	0·1405	−1	0·719	0·1405				=	0
3		0·719	−1	0·1402	0·1402			=	0·013
4		0·1645	0·1645	−1	0·342	0·329		=	0
5			0·1645	0·3147	−1	0·1643	0·329	=	0·013
6				0·284	0·142	−1	0·1477	=	423·87
7					0·284	0·1475	−1	=	564·18

(ii) *Transient heat flow FDEs.* Energy balance equations using geometric factors can be derived in the same general way as for square mesh systems. However, whereas for square meshes the area associated with each mode is a square, the areas derived from triangular meshes are usually irregular polygons. For example, the area associated with node 3 is *abcd*; the numerical value of this area can be calculated in the manner described in Section 4.2 of Chapter 4.

In the transient problem, although the temperature along the base of the fin will be assumed constant at a given time, the value will be derived by treating the plate under the fin in a simple one-dimensional manner, so that

$$T'_{11} = \mathrm{Fo}\left[T_8 + T_{12} + T_{11}\left(\frac{1}{\mathrm{Fo}} - 2\right)\right]$$

and

$$T'_8 = \mathrm{Fo}\left[T_6 + T_{11} + T_8\left(\frac{1}{\mathrm{Fo}} - 2\right)\right]$$

Using the numerical data supplied in the earlier part of this problem, the Fourier number Fo is equal to $1{\cdot}0\Delta t$. The lower surface, represented by T_{12}, will be assumed to take up a temperature of 505 °C at $t = 0$, and 1000 °C thereafter.

The FDEs for the remainder of the nodes can be derived from the energy balance, as for example for node 3 of Fig. 8.24:

$$F_{13} = 0$$
$$F_{23} = 1{\cdot}6$$
$$F_{43} = 0{\cdot}3125$$
$$F_{53} = 0{\cdot}3125$$

The net sum of the heat flows to node 3 by conduction and convection will change the temperature from T_3 to T'_3 in a time t, *i.e.*

$$k[F_{13}(T_1 - T_3) + F_{23}(T_2 - T_3) + F_{43}(T_4 - T_3) + F_{53}(T_5 - T_3)] + hA(T_A - T_3) = (\text{mass of segment } abcd) \times C_p(T'_3 - T_3)/\Delta t$$

Inserting the relevant numerical values gives the FDE for T'_3:

$$T'_3 = \Delta t\left[1{\cdot}698T_1 + 4{\cdot}348T_2 + 0{\cdot}849(T_4 - T_5) + T_3\left(\frac{1}{\Delta t} - 7{\cdot}75\right) + 0{\cdot}1475\right]$$

The energy balance method of deriving this sort of FDE evolves what is termed an explicit equation in which a temperature at a time Δt in the future is predicted from temperatures already known at the present time. These equations are always stable if the value of Δt is so chosen as to avoid negative

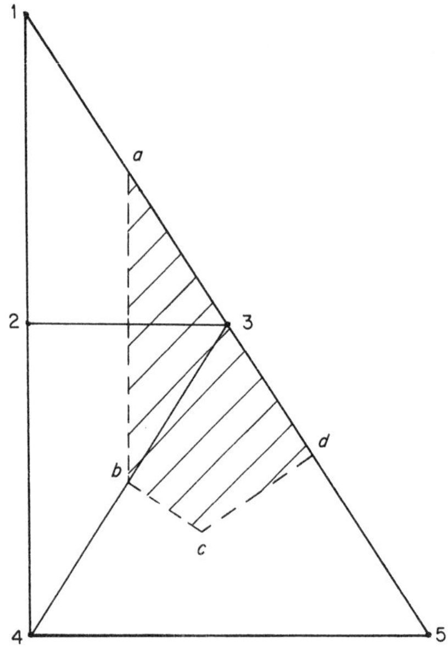

FIG. 8.24. Typical surface node.

coefficients in the FDE. In the equation for T_3', for example, this would require Δt to be not greater than (1/7·75) s. The FDE for each node will require a different Δt for stability to be guaranteed. Thus, for the whole equation set to be stable, the least value of Δt must be discovered and then used for every equation.

The whole equation set is as follows:

Node	FDE
1	$T_1' = \Delta t\left[4{\cdot}0T_2 + 0{\cdot}188 + T_1\left(\frac{1}{\Delta t} - 4{\cdot}0188\right)\right]$
2	$T_2' = \Delta t\left[5{\cdot}12T_3 + T_1 + T_4 + T_2\left(\frac{1}{\Delta t} - 7{\cdot}12\right)\right]$
3	$T_3' = \Delta t\left[1{\cdot}698T_1 + 4{\cdot}348T_2 + 0{\cdot}849(T_4 + T_5) + T_3\left(\frac{1}{\Delta t} - 7{\cdot}75\right) + 0{\cdot}1475\right]$

4 $$T_4' = \Delta t\left[0{\cdot}528(T_2 + T_3) + 1{\cdot}056T_6 + 1{\cdot}098T_5 + T_4\left(\frac{1}{\Delta t} - 3{\cdot}21\right)\right]$$

5 $$T_5' = \Delta t\left[0{\cdot}408(T_3 + T_6) + 0{\cdot}816T_7 + 0{\cdot}849T_4 + T_5\left(\frac{1}{\Delta t} - 2{\cdot}496\right) + 0{\cdot}1475\right]$$

6 $$T_6' = \Delta t\left[0{\cdot}701T_4 + 0{\cdot}350T_5 + 0{\cdot}365T_7 + 1{\cdot}051T_8 + T_6\left(\frac{1}{\Delta t} - 2{\cdot}466\right)\right]$$

7 $$T_7' = \Delta t\left[0{\cdot}762T_5 + 0{\cdot}396T_6 + 1{\cdot}524T_8 + T_7\left(\frac{1}{\Delta t} - 2{\cdot}697\right) + 0{\cdot}1495\right]$$

8 $$T_8' = \Delta t\left[T_6 + T_{11} + T_8\left(\frac{1}{\Delta t} - 2{\cdot}0\right)\right]$$

11 $$T_{11}' = \Delta t\left[T_8 + T_{12} + T_{11}\left(\frac{1}{\Delta t} - 2{\cdot}0\right)\right]$$

The most critical stability requirement (smallest Δt) is produced by the FDE for node 3, that is, Δt must not be greater than 0·129 s.

8.5.4. Choice of Solution Procedure

The equation set in the steady-state problem is best solved by a direct method such as Gauss elimination (see Section 7.2 of Chapter 7), and Program 2.1 of Chapter 7 solves equations in this manner.

The transient problem evolves an equation set which demands an explicit method of solution (see Section 7.5 of Chapter 7 and Program 5.1). The method consists of beginning the calculation with every temperature throughout the fin already known, and successively applying the FDEs until the number of time-steps reaches the specified value. For this particular problem, the total time of interest is 30 s, which will require 232 cycles of calculation if the critical value of Δt (= 0·129 s) is employed.

8.5.5. Computed Results and Discussion

The computed steady-state and transient results are shown in Table 8.3 for nodes 1 to 8 inclusive. The steady-state results, though admittedly based on a coarse mesh, permit a rough calculation of the fin efficiency. The

TABLE 8.3

Node	*Steady-state temperature* (°C)	*Temperature after* 30 *s* (°C)
1	932·6	915·9
2	937·0	921·2
3 and 3′	936·4	920·9
4	944·3	932·2
5 and 5′	929·4	929·0
6	967·5	944·4
7 and 7′	970·9	943·2
10′, 9′, 8, 9 and 10	995·0	961·0
11	997·5	979·9
12	1000·0	1000·0

average fin temperature under steady-state conditions is 950·1 °C. For heat flow through the fin:

$$Q = hA(T_{\text{base of fin}} - T_A) \times \text{Fin efficiency}$$

Also

$$Q = hA(T_{\text{fin average}} - T_A)$$

and hence, inserting numerical values, the fin efficiency

$$= \frac{(950{\cdot}1 - 10)}{(995{\cdot}0 - 10)} = 95{\cdot}4\,\%$$

8.6. HEAT FLOW INVOLVING CHANGE OF PHASE

8.6.1. Problem Statement

During the heating or cooling of steel slabs through the austenite–ferrite transformation-temperature range a peak or plateau can be observed in the rate at which the temperature changes. This is caused by the change in free energy associated with transformation of austenite to ferrite, resulting in a liberation of energy, during a cooling process, or the transformation of ferrite to austenite, resulting in an absorption of energy, during a heating process.

This particular problem is concerned with transient heating of a simple steel slab, and the effects of the phase change on the temperature distributions throughout the material. Normally the boundary conditions would be specified in terms of radiative and convective coefficients as in

Davis (1972), but here the boundary temperature itself will be specified so as to concentrate attention on the point of the problem.

The thermal properties are evaluated at a mean temperature of 500 °C; these are taken as:

Thermal conductivity	= 50 W/mK
Specific heat	= 650 J/kgK
Density	= 7700 kg/m^3
Free energy absorbed during transformation (this change is taken to occur at about 730 °C)	= 56 kJ/kg

The thickness of the slab is 100 mm, and the other dimensions are large compared with this; the problem will therefore be considered as one-dimensional.

8.6.2. Type and Size of Grid

The grid will comprise a series of parallel lines. Starting with a coarse grid, a convenient grid size is 12·5 mm, which generates eight segments within the slab.

8.6.3. Development of Finite Difference Equations

The differential equation for one-dimensional transient heat flow with internal heat generation is

$$k\frac{\partial^2 T}{\partial x^2} + H = \rho C_\mathrm{p}\frac{\partial T}{\partial t}$$

for temperature-independent thermal properties.

The internal heat generation term in this particular problem is not a constant to be applied at all times. It is, in fact, a quantity (rather than a rate of flow) of heat to be absorbed when the temperature reaches 730 °C. The simplest way to include this amount of heat in the calculation is as follows:

(i) Calculate the temperature fall that would occur if the latent heat (56 kJ/kg) were abstracted from unit mass of material of specific heat equal to 700 J/kg:

$$\text{Temperature fall} = \frac{\text{Latent heat}}{\text{Specific heat}} = 80\,^\circ\text{C}$$

(ii) During the application of the FDEs, whenever a node temperature exceeds the transformation temperature (730 °C), the temperature should be changed to a value of 730 °C, and the difference between the predicted temperature and 730 ° duly noted. During subsequent iterations this adjustment procedure should continue until the cumulative total of the adjustments is equal to 80 °C. At that point the node temperature should be allowed to increase in the normal manner.

The FDE for any internal node will therefore not include the internal heat generation term, and will be of the form

$$k\left(\frac{T_{i-1} - 2T_i + T_{i+1}}{(\Delta x)^2}\right) = \rho C_p\left(\frac{T_1' - T_i}{\Delta t}\right)$$

where T_{i-1}, T_i and T_{i+1} are temperatures at adjacent nodes. The equation resolves to

$$T_1' = \text{Fo}\left[T_{i-1} + T_{i+1} + T_i\left(\frac{1}{\text{Fo}} - 2\right)\right] \quad \text{where Fo} = \frac{k\Delta t}{\rho C_p(\Delta x)^2}$$

The above equation has been derived explicitly and is therefore subject to the stability requirement of avoiding negative coefficients. For a Fourier number of 0·5, the critical value of the time-step is 7·82 s. The actual value of the time-step is not of particular interest here, as enough iterations will be carried out to illustrate the effect of the latent heat.

8.6.4. Choice of Solution Procedure

The method of derivation of the FDEs demands an explicit type of solution procedure, in which the temperatures are specified at the beginning of the heating process, and then the FDEs progressively applied across the mesh. The program will be of the type discussed in Section 7.5 of Chapter 7 (see Program 5.1).

8.6.5. Computed Results and Discussion

As the steel slab was subjected to symmetrical boundary conditions, it meant that only half the material needed to be analysed. Figure 8.25 shows the predicted temperature variations at two sections within the material. The energy adsorption at 730 °C is clear to see, particularly when the results are compared with predictions which ignore the latent heat effect. Figure 8.26 shows the predicted temperature distribution across the section after 3·25 min, again illustrating the effect of latent heat. The difference in energy stored, represented by the two curves, will of course be exactly equal to the amount of latent heat associated with this particular mass of metal.

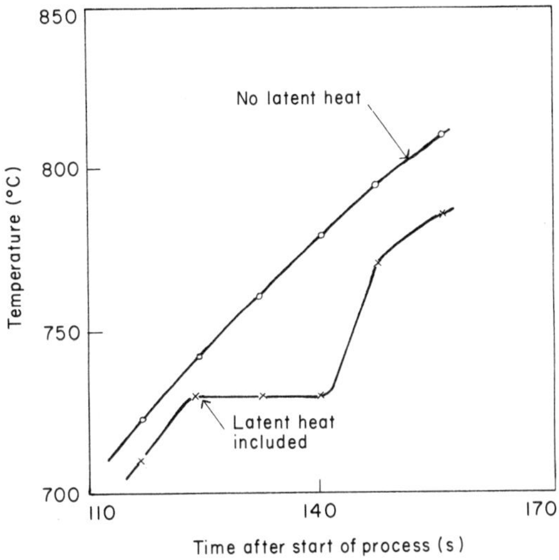

FIG. 8.25. Temperature variation at a section 25 mm from surface.

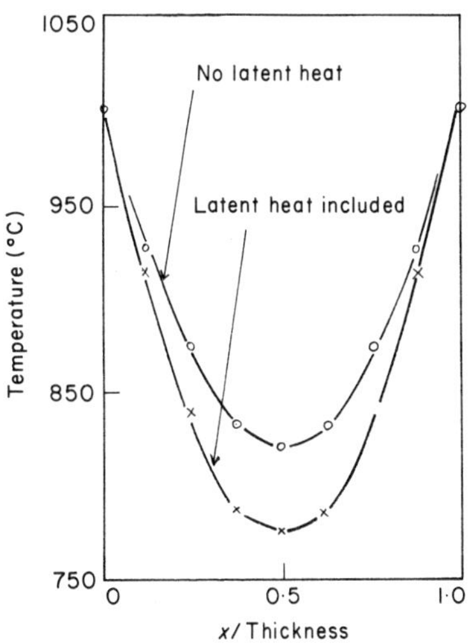

FIG. 8.26. Temperature distribution through steel slab after 3·25 min.

8.7. COMPARISON OF SOLUTION TECHNIQUES FOR A 1-D TRANSIENT PROBLEM: BY EXPLICIT TECHNIQUE AND THE IMPLICIT CRANK–NICOLSON APPROACH

The problem has been stated already in Section 7.5 of Chapter 7 (Program 5.2). It concerns a bar of 1 m length which is cooled at both ends. The following data apply, and the solutions will be compared at the time 907·8 s:

Initial temperatures in bar: varies linearly from 50 °C at left-hand end to 100 °C at right-hand end
Thermal properties of bar material: conductivity = 400 W/mK, specific heat = 495 J/kg K
Density: 8892 kg/m^3
Surface heat transfer coefficient (at both ends): 1600 W/m^{2}K
Temperature of coolant: 20 °C

8.7.1. Explicit Solution

The energy balance approach to developing the FDE is used (Section 3.5 of Chapter 3), with the usual notation:

$$\text{Internal node:} \quad T_{n,j+1} = \text{Fo}\left[T_{n-1,j} + T_{n+1,j} + T_{n,j}\left(\frac{1}{\text{Fo}} - 1\right)\right]$$

$$\text{Boundary node:} \quad T_{0,j+1} = 2\text{Fo}\left[\text{Bi} \cdot T_A + T_{1,j} + T_{0,j}\left(\frac{1}{2\text{Fo}} - 1 - \text{Bi}\right)\right]$$

The stability criterion (see Section 4.5 of Chapter 4) requires that these explicit equations will not contain any negative coefficients. The FDE for the boundary node determines the minimum value of the Fourier number permissible, that is,

$$\frac{1}{2\text{Fo}} \geq 1 + \text{Bi}$$

Also

$$\text{Fo} = \frac{\alpha \cdot \Delta t}{(\Delta x)^2} = 9{\cdot}08 \times 10^{-5}\frac{\Delta t}{(\Delta x)^2}$$

$$\text{Bi} = \frac{h\Delta x}{k} = 4 \cdot \Delta x$$

The total time for the solution is 907·8 s. There is a range of values for Δx and Fo which will satisfy all constraints. If the bar is subdivided into seven slices and Fo is taken as 0·25, then the problem can be handled by a desk calculator in about 1 h. Since $\Delta x = (\frac{1}{7})$ m, the Biot number Bi $= \frac{4}{7}$ and Fo $\leq 1/[2(1 + 4/7)]$ for stability. Thus, we require Fo $\leq 0{\cdot}318$. So the value of Fo $= 0{\cdot}25$ (chosen, like Fo $= 0{\cdot}5$ or $0{\cdot}333$, for ease of desk calculation of the internal node value calculations) will satisfy the stability criteria.

The FDEs are:

$$\text{Internal node:} \quad T_{n,j+1} = 0{\cdot}25[T_{n-1,j} + T_{n+1,j} + 2T_{n,j}]$$

$$\text{Two boundary nodes:} \quad T_{0,j+1} = 5{\cdot}714 + 0{\cdot}5T_{1,j} + 0{\cdot}214T_{0,j}$$

and the time-step is

$$\Delta t = \frac{\text{Fo}\,(\Delta x)^2}{\alpha} = \frac{0{\cdot}25 \times 1/49}{9{\cdot}08 \times 10^{-5}} = 56{\cdot}19\,\text{s}$$

$\therefore$ Number of time-steps required for $908{\cdot}7$ s $= 16$.

The calculation can now be evolved as shown in Table 8.4.

TABLE 8.4
EXPLICIT SOLUTION OF 1-D TRANSIENT PROBLEM

T	T_A	*LH end* T_0	T_1	T_2	T_3	T_4	T_5	T_6	*RH end* T_7
0	20	50	57·14	64·29	71·43	78·57	85·714	92·86	100
1	20	45·12	57·14	64·29	71·43	78·57	85·714	92·86	73·54
2	20	43·94	55·92	64·29	71·43	78·57	85·714	86·24	67·88
3	20	43·07	55·02	64·23	71·43	78·57	84·06	81·52	63·36
·									
·									
·									
16	20	38·74	48·90	57·23	62·83	64·73	62·22	54·85	43·12
Distance along bar		0	0·143	0·286	0·429	0·571	0·714	0·857	1·00

8.7.2. Implicit Crank–Nicolson Solution

This example is discussed and the FDEs given in Section 7.5 of Chapter 7 (see Program 5.2).

Left-hand end: Using a value of $\Delta x = \frac{1}{20}$ (*i.e.* 20 segments), Fo $= 1$, and $h/k = 4$:

$$-4{\cdot}4T_{0,j+1} = -2T_{1,j+1} - 2T_{1,j} + 0{\cdot}4T_{0,j} - 16$$

Right-hand end:

$$-4{\cdot}4T_{n,j+1} = -2T_{n-1,j+1} - 2T_{n-1,j} + 0{\cdot}4T_{n,j} - 16$$

Interior nodes:

$$T_{i,j+1} = \tfrac{1}{4}[T_{i-1,j+1} + T_{i-1,j} + T_{i+1,j+1} + T_{i+1,j}]$$

Since

$$\Delta t = \frac{\text{Fo}\,(\Delta x)^2}{\alpha}, \quad \Delta t = \frac{1 \times 1/400}{9{\cdot}08 \times 10^{-5}} = 27{\cdot}53$$

Hence the number of time-steps required is 33.

These FDEs form a tridiagonal system and may be solved by Program 2.2 (see Section 7.2 of Chapter 7) via simple Gaussian elimination, which is incorporated into Program 5.2 of Chapter 7.

For ease of comparison, Fig. 8.27 has been drawn to show both predictions. The final predictions are seen to be virtually identical. This agreement is expected and should not hide the following points:

(i) The implicit solution uses 33 time-steps and a subdivision of 20 segments. Using 20 segments with the explicit method requires that $\text{Fo} \leq 0{\cdot}416$. This means that at least 79 time-steps will be needed to ensure stability.

(ii) If the number of time-steps required is excessive, then the implicit

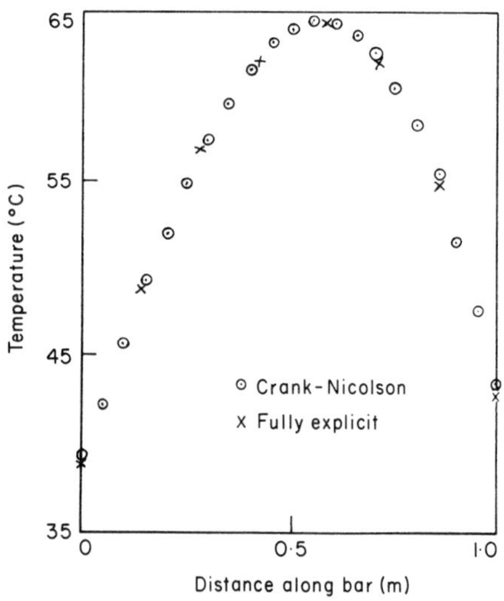

FIG. 8.27. Temperature profiles generated by different solution techniques.

technique permits alteration of the Fo to increase the value of Δt without any change in Δx. The explicit method is limited by the stability requirement, so that an increase in Δt must be accompanied by an increase in Δx ($\Delta x \propto (\Delta t)^{1/2}$) for constant Fo. Increasing these will of course reduce the accuracy of solution.

8.8. TWO-DIMENSIONAL TRANSIENT BEHAVIOUR: AN ILLUSTRATION OF THE ALTERNATING DIRECTION IMPLICIT (ADI) SOLUTION PROCEDURE

8.8.1. Problem Statement

The problem has been chosen primarily as a vehicle to demonstrate a particular solution procedure and is therefore somewhat simplified to permit emphasis on the procedure rather than the problem.

A metal bar of rectangular section 40 mm by 80 mm, initially at a temperature of 600 °C throughout, is plunged into a water bath and the problem is to determine the temperature distribution through the bar after a period of, say, 15 s. The third dimension of the bar is considered large compared to the other two and therefore the problem will be taken as essentially two-dimensional.

Since the water adjacent to the metal will immediately boil, the initial heat transfer coefficient will be of such a high order that the temperature difference between the surface of the metal and the water will be so low as to justify the proposition that the surface temperature be taken as 100 °C.

8.8.2. Type and Size of Grid

Since the primary purpose of this example is to illustrate the solution procedure, a coarse square grid of side 10 mm is used. As the section is symmetrical, only a segment (shown shaded in Fig. 8.28) need be used. Thermal properties are taken as:

Specific heat	= 750 J/kgK
Thermal conductivity	= 60 W/mK
Density	= 8950 kg/m^3

8.8.3. Development of Finite Difference Equations

The FDEs are, as before, developed from energy balance equations, but in such a form as to suit the ADI solution procedure. This procedure has been discussed previously in Section 7.5 of Chapter 7, but to repeat briefly: the value of the ADI method is that there is no restriction on the numerical value of the time-step Δt for stability. This is not the case for explicit FDEs, where the requirement of avoiding negative coefficients limits the allowable value for Δt, thereby often requiring a large number of calculations to be made.

The ADI method requires two sets of FDEs to be derived for a given time-step. The equations are a mixture of explicit and implicit terms: for the first half of the time-step, the FDEs for the x-direction terms are implicit in form and for the y-direction explicit, and for the second half of the time-step vice versa. Since the temperature at the boundary is known, only FDEs for the internal nodes need to be derived.

The equations for the first half of the time-step, by energy balance method in usual notation, are, $\Delta = \Delta x = \Delta y$:

Net heat flow to node (I, J) = 'Mass' of node $\times$ Specific heat $\times$ Rate of temperature change

which is

$$\left(\frac{kA}{\Delta}(T'_{I,J-1} - T'_{I,J}) + \frac{kA}{\Delta}(T'_{I,J+1} - T'_{I,J}) + \frac{kA}{\Delta}(T_{I-1,J} - T_{I,J}) + \frac{kA}{\Delta}(T_{I+1,J} - T_{I,J})\right) = \rho(A\Delta)C_{\mathrm{p}}\frac{(T'_{I,J} - T_{I,J})}{(\Delta t/2)}$$

Similarly, the equation for the second half of the time-step is

$$\left(\frac{kA}{\Delta}\right)\left((T_{I,J-1} - T_{I,J}) + (T_{I,J+1} - T_{I,J}) + (T'_{I-1,J} - T'_{I,J}) + (T'_{I+1,J} - T'_{I,J})\right) = 2\rho A\Delta C_{\mathrm{p}}(T'_{I,J} - T_{I,J})/\Delta t$$

The equations are so written that the left-hand side of the equation comprises 'unknown' values, and the right-hand side already known values. Represented in matrix form, the equations are seen to be of the same type as the 1-D elliptic (steady-state) equations already discussed in Chapter 5, Section 5.1, and Chapter 7, Sections 7.2 and 7.5, and are therefore suitable to be solved by a direct method such as Gauss elimination (Program 2.1 or preferably 2.2—see Section 7.2 of Chapter 7). Thus we have, where the Fourier number $\mathrm{Fo} = \dfrac{k}{2\rho C_{\mathrm{p}}}\dfrac{\Delta t}{\Delta^2}$:

For the first half time-step (x-direction implicit, y-direction explicit):

Node	T_1'	T_2'	T_3'	T_4'	T_5'	T_6'	T_7'	T_8'	
1	$\left(-\frac{1}{Fo}-2\right)$	1							$= -200 + \left(2-\frac{1}{Fo}\right)T_1 - T_5$
2	1	$\left(-\frac{1}{Fo}-2\right)$	1						$= -100 + \left(2-\frac{1}{Fo}\right)T_2 - T_6$
3		1	$\left(-\frac{1}{Fo}-2\right)$	1					$= -100 + \left(2-\frac{1}{Fo}\right)T_3 - T_7$
4			2	$\left(-\frac{1}{Fo}-2\right)$	0				$= -100 + \left(2-\frac{1}{Fo}\right)T_4 - T_8$
5					$\left(-2-\frac{1}{Fo}\right)$	1			$= -100 + \left(2-\frac{1}{Fo}\right)T_5 - 2T_1$
6					1	$\left(-\frac{1}{Fo}-2\right)$	1		$= \left(2-\frac{1}{Fo}\right)T_6 - 2T_2$
7						1	$\left(-\frac{1}{Fo}-2\right)$	1	$= \left(2-\frac{1}{Fo}\right)T_7 - 2T_3$
8							2	$\left(-\frac{1}{Fo}-2\right)$	$= \left(2-\frac{1}{Fo}\right)T_8 - 2T_4$

For the second half time-step (x-direction explicitly, y-direction implicit):

Node	T'_1	T'_2	T'_3	T'_4	T'_5	T'_6	T'_7	T'_8	
1	$\left(-\frac{1}{\mathrm{Fo}}-2\right)$				1				$= -200 + \left(2 - \frac{1}{\mathrm{Fo}}\right)T_1 - T_2$
2		$\left(-\frac{1}{\mathrm{Fo}}-2\right)$				1			$= -100 - T_1 + \left(2 - \frac{1}{\mathrm{Fo}}\right)T_2 - T_3$
3			$\left(-\frac{1}{\mathrm{Fo}}-2\right)$				1		$= -100 - T_2 + \left(2 - \frac{1}{\mathrm{Fo}}\right)T_3 - T_4$
4				$\left(-\frac{1}{\mathrm{Fo}}-2\right)$				1	$= -100 + \left(2 - \frac{1}{\mathrm{Fo}}\right)T_4 - 2T_3$
5	2				$\left(-\frac{1}{\mathrm{Fo}}-2\right)$				$= -100 + \left(2 - \frac{1}{\mathrm{Fo}}\right)T_5 - T_6$
6		2				$\left(-\frac{1}{\mathrm{Fo}}-2\right)$			$= -T_5 + \left(2 - \frac{1}{\mathrm{Fo}}\right)T_6 - T_7$
7			2				$\left(-\frac{1}{\mathrm{Fo}}-2\right)$		$= -T_6 + \left(2 - \frac{1}{\mathrm{Fo}}\right)T_7 - T_8$
8				2				$\left(-\frac{1}{\mathrm{Fo}}-2\right)$	$= -2T_7 + \left(2 - \frac{1}{\mathrm{Fo}}\right)T_8$

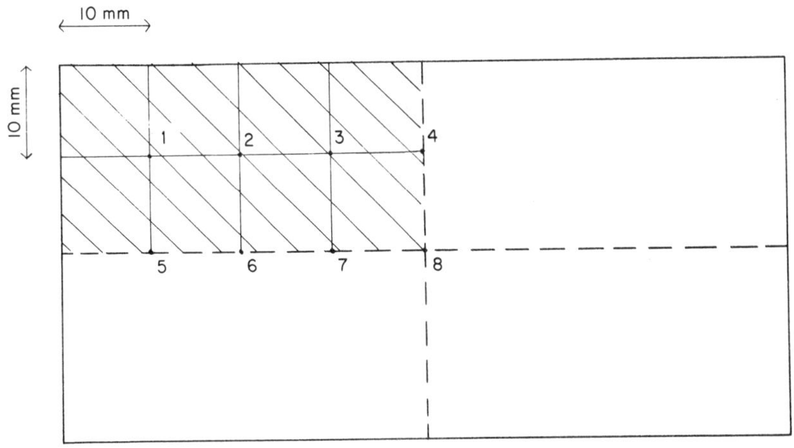

FIG. 8.28. Quarter section of rectangular metal bar (shaded).

8.8.4. Choice of Solution Procedure

The presentation of the equations in matrix form shows the tridiagonal nature of the set (reordering the second set in the order 1, 5, 2, 6, . . ., first) and suggests that Program 2.2 (see Section .7.2 of Chapter 7) is the most appropriate type of solution procedure. The first equation set is solved, and the solutions are then used in the second equations, leading to the final values at the end of the whole time-step Δt. The Fourier number Fo is initially taken as unity—a value four times that permitted if the problem had been solved by an explicit procedure. This requires five time-steps as against a minimum of 20 for stability with the explicit method.

8.8.5. Computed Results and Discussion

The computed results are shown in Fig. 8.29 with the predictions using an explicit solution shown in brackets.

It should be remembered that the number of time-steps taken is four times less than the explicit case. The results by the two procedures obviously compare very well.

If a less coarse grid had been chosen (say of 5 mm square), the required number of time-steps using an explicit method would be 80 (taking the critical Fourier number of 0·25). It would be quite permissible with the ADI method to use a Fourier number of 2 and thereby reduce the required number of time-steps to ten. Of course, the accuracy of the solution will diminish as the number of time-steps is reduced, so the reader should be careful in using too great a value for Fo.

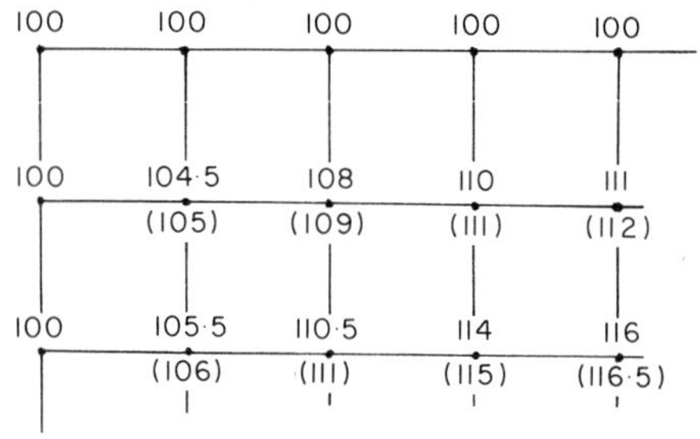

FIG. 8.29. Computed results after 15 s by ADI method (with explicit method results in brackets).

8.9. HEAT FLOW THROUGH REFRACTORY WALLS COMPRISING DIFFERENT MATERIALS WITH TEMPERATURE-DEPENDENT PROPERTIES

8.9.1. Problem Statement

In the past it has been the practice with refractory-lined structures to construct the refractory wall from two or three layers of brick, each layer having different thermal properties so as to minimise the heat loss from the structure and also to retain as much heat as possible within the refractory under all working conditions.

The same objectives are now achieved by using a single refractory material, termed a 'monolithic' refractory. This material is cast *in situ* to the required thickness (in the same way as concrete) and has the extra advantage that repairs to the inner refractory can be 'sprayed' on to the damaged area without the need for the slow business of 're-bricking', as previously applied.

When damage to the lining does occur, the effective local thickness of the lining is reduced and so the temperature of the metal shell which surrounds the refractory is higher than normal. In practice, the effect of local 'hot-spots' on the metal shell is offset by cooling the shell with water. Even so, there is a critical effective thickness of refractory below which the plant has to be closed down and repairs made.

The problems, then, are these:

(i) for the steady state, to decide what is the magnitude of the cooling effect due to the water (that is, what heat transfer coefficient applies), and secondly, to relate measurements of outside shell temperatures with refractory thicknesses;*

(ii) for the transient condition, to predict the thermal response of the composite structure when the water cooling ceases, air cooling takes over, and then subsequently the water cooling is returned.

8.9.2. Thermal Properties

For the refractory material:

$$\text{Specific heat} = 780\ \text{J/kgK}$$
$$\text{Density} = 1811\ \text{kg/m}^3$$

The thermal conductivity must be expressed as a polynomial in temperature:

$$k_R = (1{\cdot}11 - 2{\cdot}1 \times 10^{-4}T + 4{\cdot}64 \times 10^{-8}T^2)\ \text{W/mK}$$

where T is the temperature in °C. For the metal:

$$\text{Specific heat} = 500\ \text{J/kgK}$$
$$\text{Density} = 8000\ \text{kg/m}^3$$
$$\text{Thermal conductivity} = 45\ \text{W/mK}$$

The refractory-lined structure for this analysis will be taken to be a large blastfurnace. The structure is of circular cross-section but the diameter and length are of the order of metres whereas the total maximum thickness of lining plus metal shell is only 320 mm. The problem will therefore be treated as one-dimensional, and although this is a simple model, it serves to illustrate the finite difference treatment of variable thermal properties and heat flow through composite structures.

8.9.3. Type and Size of Grid

There is little to decide on this matter as the problem has been simplified to one-dimensional. A grid comprising parallel lines a distance 40 mm apart will give adequate coverage of both refractory and shell material.

8.9.4. Development of Finite Difference Equations (Steady-State)

Since the inside surface temperature of the refractory is specified at 1300 °C, there are only four typical FDEs to derive: one each for the internal nodes of the refractory and the metal shell, one for the interface between the two materials, and one for the outer node on the boundary of the metal shell.

* Personal correspondence (1973) to D. R. Croft from Thermographic Section, British Steel Corporation, Sketty Hall, Swansea.

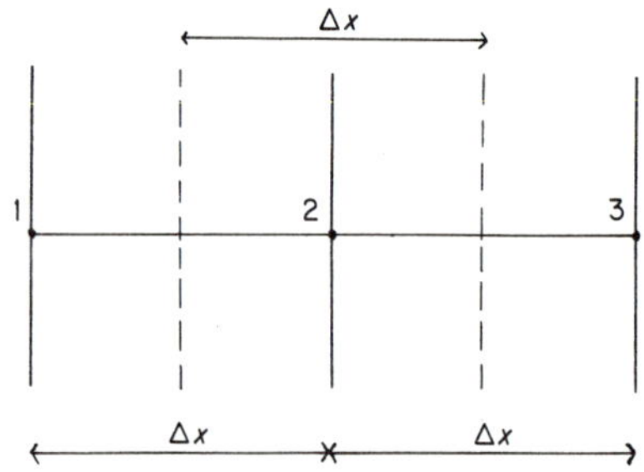

FIG. 8.30. Typical internal node 2.

(i) *FDE for an internal node in the refractory.* Using the energy balance concept, the net heat flow to node 2 of Fig. 8.30 will be zero, that is,

$$\frac{k_{12}A(T_1 - T_2)}{\Delta x} + \frac{k_{32}A(T_3 - T_2)}{\Delta x} = 0$$

where k_{12} is evaluated at $(T_1 + T_2)/2$, and k_{32} is evaluated at $(T_3 + T_2)/2$. Thus, since $A = \Delta x \cdot 1$,

$$T_2 = \frac{(k_{12}T_1 + k_{32}T_3)}{(k_{12} + k_{32})}$$

Since the thermal conductivity of the refractory is a polynomial function of temperature, the above equation will not be linear.

(ii) *FDE for internal node in the metal shell.* Since the metal is taken to have a constant thermal conductivity, the heat flow will be governed by the simple equation

$$\frac{\mathrm{d}^2T}{\mathrm{d}x^2} = 0$$

For a given node i

$$T_i = \frac{(T_{i-1} + T_{i+1})}{2}$$

(iii) *FDE for the interface node.* The difficulty here is that the node represents two different materials having different thermal properties. A sketch of the area surrounding the interface is shown in Fig. 8.31. Again using the energy balance concept, this states that the net flow to node I is zero in the steady state. Thus

$$\frac{k_{RI}A(T_R - T_I)}{\Delta x} + \frac{k_M A(T_M - T_I)}{\Delta x} = 0$$

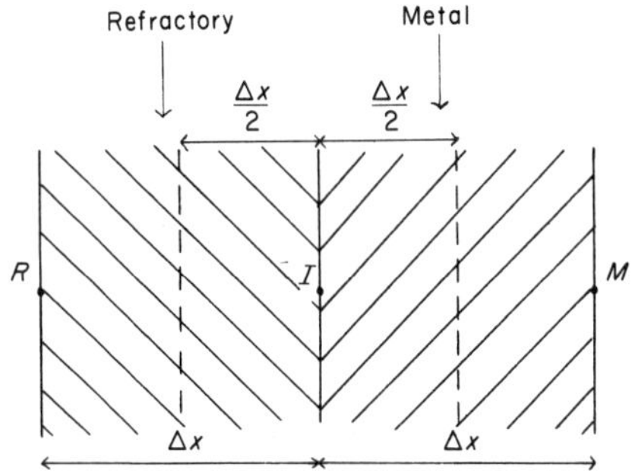

FIG. 8.31. Typical interface node *I*.

giving

$$T_I = \frac{(k_{RI}T_R + k_M T_M)}{(k_{RI} + k_M)}$$

where k_{RI} is evaluated from the polynomial at a temperature $(T_R + T_I)/2$ and k_M is a constant.

(iv) *FDE for the boundary node*. A sketch of the region surrounding a typical boundary node *B* is shown in Fig. 8.32. Provision will be made in the FDEs for both convective and radiative heat flow from the boundary. Applying an energy balance to the boundary node *B*:

$$\frac{k_M A(T_1 - T_B)}{\Delta x} + Q = 0$$

where $Q = hA(T_A - T_B)$ for purely convective boundary conditions, and $Q = hA(T_A - T_B) + \varepsilon\sigma A(T_A^4 - T_B^4)$ for convective and radiative boundary conditions. In this solution the emissivity is taken as 0·9.

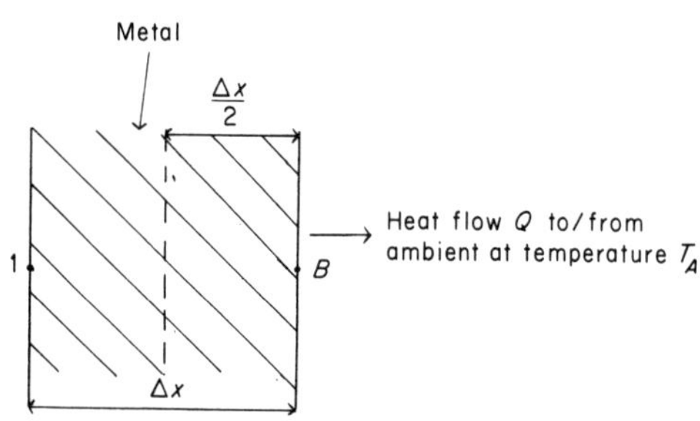

FIG. 8.32. Typical boundary node *B*.

8.9.5. Choice of Solution Procedure (Steady-State)

Apart from the equation for the internal node for the metal shell, all the equations are non-linear and therefore not eligible to be solved by a direct method. The alternative is to use an iterative procedure such as the Gauss–Seidel (see Sections 7.3 and 7.4 of Chapter 7 and Programs 3.1 and 4.1). The procedure consists of three main stages:

(i) An initial estimate of the temperature at every node must be made to establish a starting-point for the iterative process. The inside refractory temperature is fixed at 1300 °C and this value is the easiest to insert as the initial value for the other nodes in the refractory. Strictly speaking, any value of temperature may be used to initialise the procedure, but more cycles of iteration will be required if the values are not chosen reasonably carefully.

(ii) The next stage in the procedure is to apply the FDEs in a systematic sequence across the structure, starting with the node next to the inner refractory surface.

(iii) After each cycle of iteration has been carried out, some decision has to be made as to whether or not to continue the process. This decision is based on the biggest percentage change in node temperature from one iteration to the next. In this problem, only the successive changes in the outer boundary temperatures were observed and the iteration was stopped when successive temperatures differed by 0·1 % or less.

A problem occurred during the actual computation when the program was run with convective and radiative conditions applied to the outside of the metal shell. It transpired that the iterative procedure would not converge, this being due to the oscillatory variation of the predicted temperature at the outer boundary node. Owing to the highly non-linear equation which applies under these conditions at the boundary, it is not possible to specify the boundary temperature as a function of other variables only; the FDE is in the form of an equation with the boundary temperature (T_B) appearing on both sides of the equation. Eventually convergence was achieved by applying under-relaxation to the changes in T_B; in detail, the value of T_B from successive iterations was stored and an average of the two values then used as the node temperature for the next iteration, if of course further iteration was required.

8.9.6. Development of Finite Difference Equations (Transient)

As was the case for the steady-state problem, only four typical FDEs need to be derived. Only the interface node presents any difficulty which is not fully discussed in previous problems, so the FDEs for the other nodes will simply be stated:

(i) For the typical node within the refractory (node 2, see Fig. 8.30):

$$T_2' = \mathrm{Fo}\left[\frac{(k_{12}T_1 + k_{32}T_3)}{k_{12} + k_{32}} + T_2\left(\frac{1}{\mathrm{Fo}} - 1\right)\right]$$

where $\mathrm{Fo} = \dfrac{(k_{12} + k_{32})\,\Delta t}{\rho_R C_{pR}\,(\Delta x)^2}$

(ii) For a typical node within the metal shell (node i):

$$T_i' = \frac{1}{\mathrm{Fo}}\left[T_{i-1} + T_{i+1} + T_i\left(\frac{1}{\mathrm{Fo}} - 2\right)\right]$$

where $\mathrm{Fo} = \dfrac{k_M \Delta t}{\rho_M C_{pM}\,(\Delta x)^2}$

(iii) For the outer boundary node (node B, see Fig. 8.32):

$$T_B' = C\left[k_M T_1 + hT_A + \varepsilon\sigma T_A^4 + T_B\left(\frac{1}{C} - (k_M + h + \varepsilon\sigma T_B^3)\right)\right]$$

where $C = \dfrac{2\Delta t}{\rho_M \Delta x C_{pM}}$

(iv) For the interface node I (see Fig. 8.31), applying an energy balance to node I:

Net heat flow to node I by conduction = 'Mass' of segment × Specific heat × Rate of temperature change

The simplest way to evaluate the right-hand side of the equation is to consider the multiple (mass × specific heat) for the whole segment as comprising separate contributions from each material, *i.e.*

$$(\text{mass} \times \text{sp. heat})_{\text{segment}} = [(\text{mass} \times \text{sp. heat})_{\text{refractory}} + (\text{mass} \times \text{sp. heat})_{\text{metal}}]$$

$$= \left[\left(\rho_R A \frac{\Delta x}{2} C_{pR}\right) + \left(\rho_M A \frac{\Delta x}{2} C_{pM}\right)\right]$$

Thus, applying the energy balance concept will give

$$k_{RI} A \frac{(T_R - T_I)}{\Delta x} + k_M A \frac{(T_M - T_I)}{\Delta x} = A\left(\frac{\Delta x}{2}\right)(\rho_R C_{pR} + \rho_M C_{pM})\frac{(T_I' - T_I)}{\Delta t}$$

Hence

$$T'_I = \text{Fo}\left[\left(\frac{k_{RI}T_R + k_M T_M}{k_{RI} + k_M}\right) + T_I\left(\frac{1}{\text{Fo}} - 1\right)\right]$$

where $\text{Fo} = \dfrac{2\Delta t}{(\Delta x)^2} \dfrac{(k_{RI} + k_M)}{(\rho_R C_{pR} + \rho_M C_{pM})}$

All the transient equations have been derived explicitly and are therefore subject to the stability requirement of avoiding negative coefficients. Of the four typical equations, the one which generates the least allowable value for the time-step Δt is the one for the node within the metal shell; this equation demands that Δt shall not exceed 71 s.

8.9.7. Choice of Solution Procedure (Transient)

The equation set requires an explicit method of solution (see Section 7.5 of Chapter 7). The method consists of beginning the calculation by specifying the temperature at every node (produced by the steady-state analysis) and successively applying the FDEs until the number of time-steps

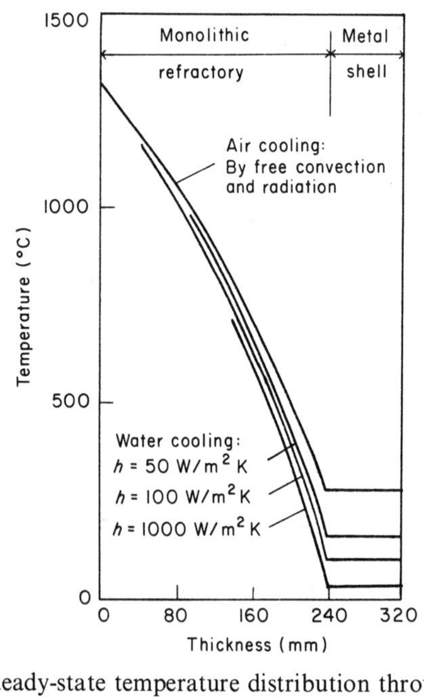

FIG. 8.33. Steady-state temperature distribution through structure.

reaches the specified value. Since the thermal properties vary at different points within the structure, the Fourier number will also vary and in turn change the coefficients in the FDEs. In fact, this does not affect the validity of the analysis since the coefficients of every FDE still add up to zero, whatever the value of Fo.

8.9.8. Computed Results and Discussion

(i) *Steady-state*. The first objective of the steady-state analysis was to discover the magnitude of the outside convective heat transfer coefficient. The steady-state program was run with various convective boundary conditions, and the resulting temperature distributions through the structure are shown in Fig. 8.33.

From thermographic measurements* the outside temperature of such a structure was known to be around 80 °C, so a coefficient of 100 W/m^2K was assumed to apply. This value was then used for the remaining calculations.

The second objective was to determine the variation of outer surface temperature with different thicknesses of refractory. Figure 8.34 shows such a variation, based on an outside coefficient of 100 W/m^2K.

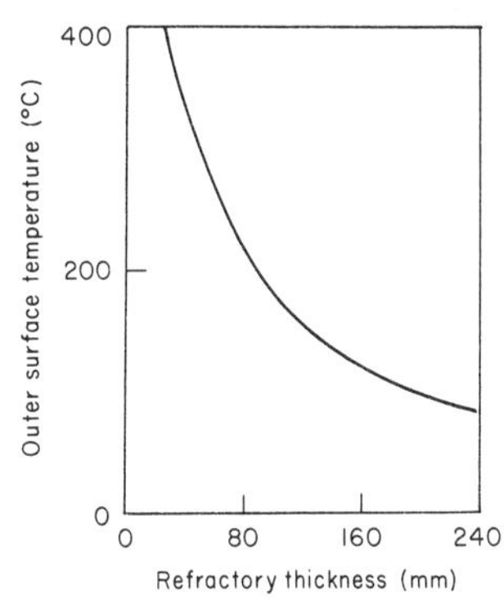

FIG. 8.34. Effect of refractory thickness on the steady-state outer surface temperature.

Although derived from a simple one-dimensional model, the results have been surprisingly useful in highlighting refractory damage from measured external temperature.*

(ii) *Transient*. The transient analysis was aimed at predicting the behaviour of the complete structure when, for whatever reason, the water

* See previous footnote, p. 233.

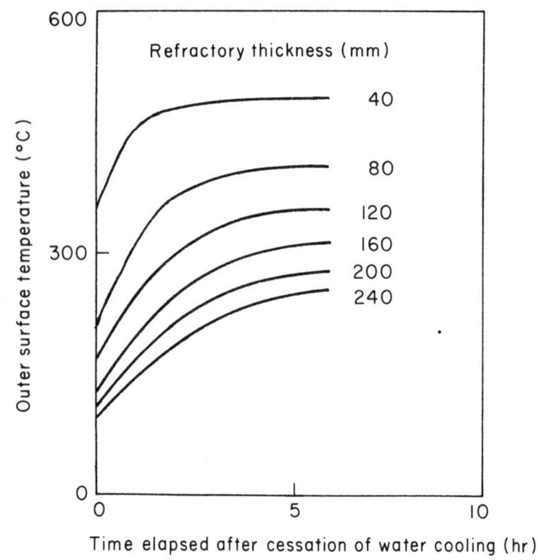

FIG. 8.35. Effect of refractory thickness on transient outer surface temperature.

cooling ceased and the shell was then cooled by free convection and radiation alone. Figure 8.35 shows the transient response of the composite structure—in terms of outside shell temperature—for a number of different refractory thicknesses.

Experimental data from a 'hot-spot' on such a structure as determined by

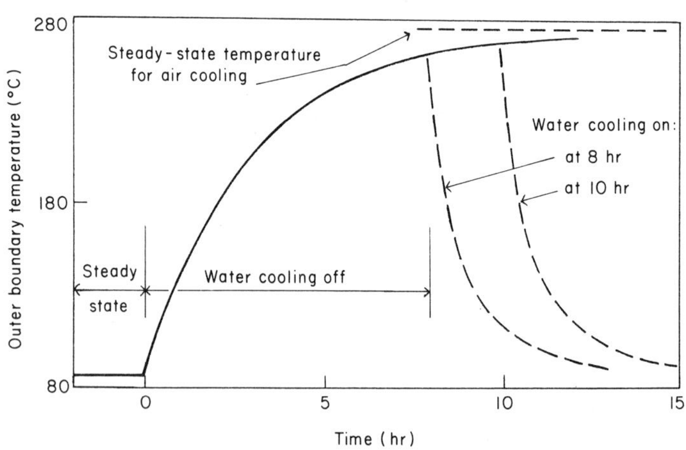

FIG. 8.36. Transient outer surface temperature when water cooling is turned off, then on again later, for 240 mm refractory thickness.

Rogers (1971) agree with the predictions for the 120 mm thickness curve. Subsequent inspection of the inside of the furnace revealed that the damaged area was reduced in thickness to around 130 mm.

Finally, the complete temperature history of the outer surface during steady-state, transient cooling, and the subsequent return to the steady state, is shown in Fig. 8.36. The figure shows the relatively rapid return to the steady state once the water cooling has been re-established. In practice, it is unlikely that this degree of thermal shock would be allowed to occur; the results at any rate provide some information on which a decision could be made.

8.10 THREE-DIMENSIONAL STEADY-STATE HEAT FLOW THROUGH A COMPOSITE REFRACTORY STRUCTURE

8.10.1. Problem Statement

In the previous problem of Section 8.9, the argument in favour of using a single (monolithic) refractory was expounded; the application does, however, pose some extra problems. In order to support the refractory, it is necessary, in some way, to 'pin' the refractory to the outer shell. This is achieved by having metallic protrusions from the metal shell into the monolithic refractory; the pins are equispaced and will be taken to be of

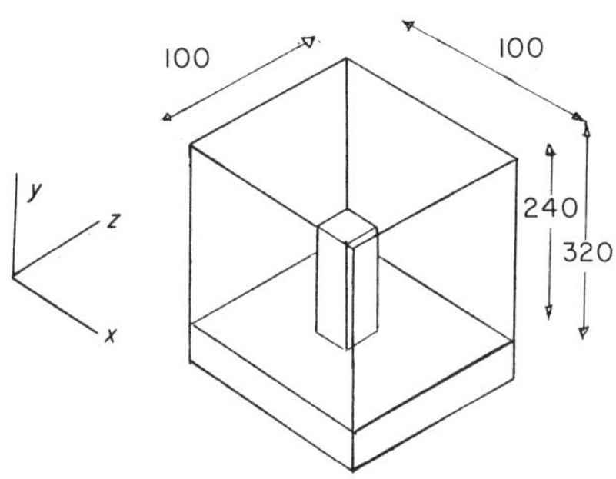

FIG. 8.37. Geometric configuration; pin dimensions are 40 × 40 × 150 mm.

square section and of the size shown in Fig. 8.37. Thermal properties are taken as for the problem of Section 8.9.

The pins obviously affect the heat flow pattern through the refractory since the metal pins increase the effective thermal conductivity of the composite wall. The problem to be solved here is to determine by how much the steady-state heat flow is increased and relate this factor to the spacing and dimensions of the pins. The configuration is clearly three-dimensional and cannot be simplified as dimensions in all three directions are of the same magnitude. There will therefore be a large number of equations and the major difficulty will be in the data handling. The equations themselves are not significantly more complex; in fact, it will be seen that the derivation and solution techniques expounded in previous examples can be quickly extended to solve this problem.

8.10.2. Type and Size of Grid

The general configuration suggests that a simple square grid would be best; all the nodes would then be representing a cubically shaped control volume. This type of mesh should make the derivation of the FDEs easier to understand and achieve. The size of the mesh should be chosen initially on the basis of 'the coarsest that can be got away with', in this case a cubic mesh of side 20 mm.

8.10.3. Development of Finite Difference Equations

For the sake of brevity it will be considered that there are three fundamentally different types of FDE to be derived: one for the nodes which are situated entirely within either the metal or refractory, one for those nodes which are situated at the interface between the two materials, and one for the boundary nodes.

(i) *FDE for a node entirely within the refractory*. As the problem is 3-D, the contribution to the energy balance will come from three directions, x, y and z, as shown in Fig. 8.38 for a typical internal node labelled 1.

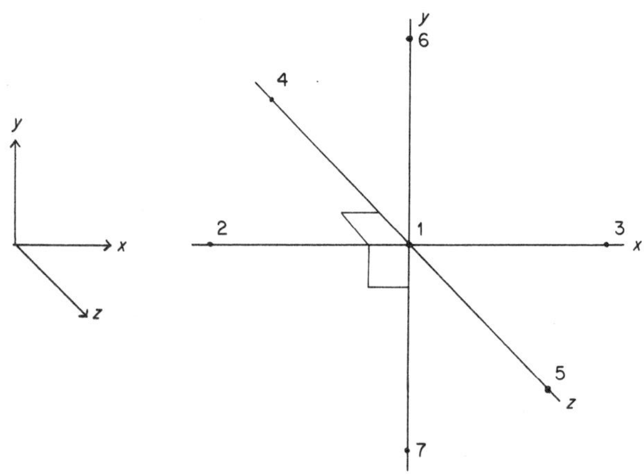

FIG. 8.38. Typical internal node 1.

The energy balance is

$$\frac{1}{\Delta x}[k_{21}A(T_2 - T_1) + k_{31}A(T_3 - T_1) + k_{41}A(T_4 - T_1) + k_{51}A(T_5 - T_1) + k_{61}A(T_6 - T_1) + k_{71}A(T_7 - T_1)] = 0$$

giving

$$T_1 = \frac{(k_{21}T_2 + k_{31}T_3 + k_{41}T_4 + k_{51}T_5 + k_{61}T_6 + k_{71}T_7)}{(k_{21} + k_{31} + k_{41} + k_{51} + k_{61} + k_{71})}$$

Remember here that the thermal conductivity of the refractory is a polynomial function of temperature, and it is better to evaluate a typical conductivity k_{21} at the average temperature rather than average its own two extreme values. Thus we use

$$k_{12} = k(T_{12}) \quad \text{where } T_{12} = \frac{(T_1 + T_2)}{2}$$

rather than

$$k_{12} = \frac{[k(T_1) + k(T_2)]}{2}$$

A similar equation will apply for all the internal nodes in the metal, except that the thermal conductivity is assumed constant. Therefore, for a typical node, the FDE will be

$$T_1 = \frac{(T_2 + T_3 + T_4 + T_5 + T_6 + T_7)}{6}$$

where the nodes are in the same geometric configuration as in Fig. 8.38.

(ii) *FDE for an interface node*. As an illustration, the FDE for one of the corner nodes (node 1 in Fig. 8.39) will be derived. Before deriving the FDEs,

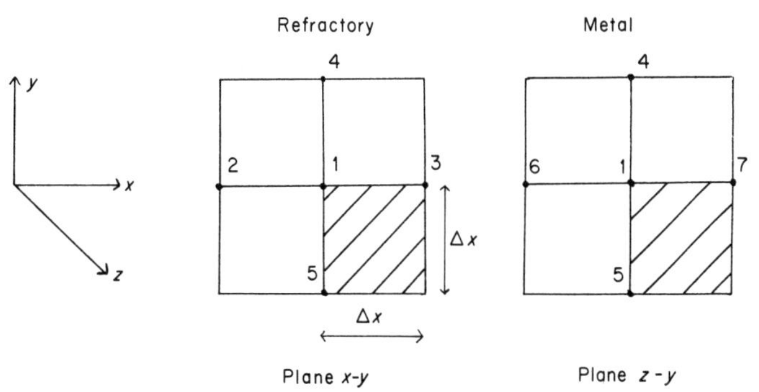

FIG. 8.39. Typical interface node 1 on an edge of pin and refractory.

it should be made clear how the heat flow from, say, node 3 to node 1 is to be described. From the figure it can be seen that half the cross-sectional area of flow is occupied by refractory and half by metal. It can quickly be shown that in this situation it is quite permissible to use an average value of the two thermal conductivities; consider two bars A and B, of exactly the same cross-sectional area A and length L, each subjected to the same temperature difference ΔT, but having different thermal conductivities k_A and k_B:

$$\text{Heat flow by conduction through bar A} = \frac{k_A A(\Delta T)}{L}$$

$$\text{Heat flow by conduction through bar B} = \frac{k_B A(\Delta T)}{L}$$

A composite bar of cross-sectional area $2A$ and length L operating between the same temperature limits will transmit a heat flow $= k_{\text{comp}} A(\Delta T)/L$. It will be seen that if the sum of the heat flows through the two bars is the same as that through the composite bar, the effective conductivity of the composite is

$$k_{\text{comp}} = \frac{(k_A + k_B)}{2}$$

Applying this concept to node 1 of Fig. 8.39, the energy balance will generate the following equation:

$$\frac{1}{\Delta x}[(k_{21}A(T_2 - T_1) + k_{31}A(T_3 - T_1) + k_{41}A(T_4 - T_1) + k_{51}A(T_5 - T_1) + k_{61}A(T_6 - T_1) + k_{71}A(T_7 - T_1)] = 0$$

where $k_{31} = (k_{\text{metal}} + k_{\text{refractory}})/2$, and the refractory conductivity is evaluated at a mean temperature equal to $(T_3 + T_1)/2$. Similar comments will apply to the evaluation of k_{51} and k_{71}. The conductances k_{21}, k_{41}, k_{61} will be evaluated from the polynomial for the refractory conductivity.

(iii) *FDE for a boundary node.* Applying the energy balance concept to node 1 of Fig. 8.40, and remembering that the cross-sectional area of flow by conduction from nodes 2, 3, 5, and 6 to node 1 is only half that for heat flow from node 4 to node 1, we have

$$\frac{1}{\Delta x}\left[k_{21}\left(\frac{A}{2}\right)(T_2 - T_1) + k_{31}\left(\frac{A}{2}\right)(T_3 - T_1) + k_{51}\left(\frac{A}{2}\right)(T_5 - T_1) + k_{61}\left(\frac{A}{2}\right)(T_6 - T_1) + k_{41}A(T_4 - T_1)\right] + hA(T_A - T_1) = 0$$

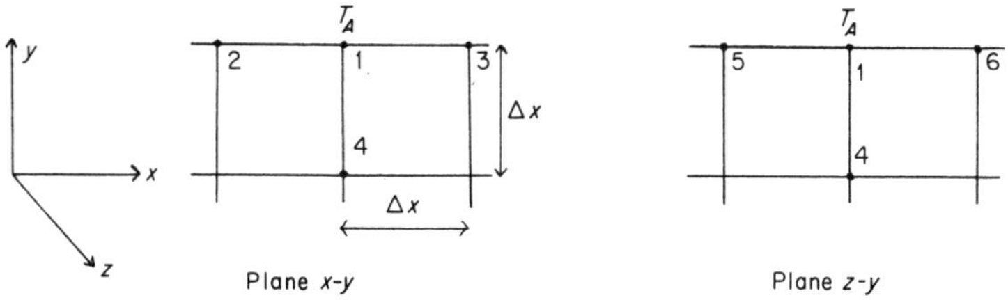

FIG. 8.40. Typical boundary node on surface of refractory exposed to ambient temperature T_A.

where $\Delta = \Delta x = \Delta y = \Delta z$ and $A = \Delta^2$. Thus

$$T_1 = \frac{k_{21} T_2 + k_{31} T_3 + k_{51} T_5 + k_{61} T_6 + 2k_{41} T_4 + 2h\Delta T_A}{(k_{21} + k_{31} + k_{51} + k_{61} + 2k_{41} + 2h\Delta)}$$

where the refractory thermal conductivities are evaluated at mean temperatures as before.

8.10.4. Choice of Solution Procedure

Even though the configuration is symmetrical (and therefore only one quarter of the object need be analysed), there are still over 600 FDEs to be solved. The non-linear nature of the equations—the k's in the coefficient are functions of the unknown temperatures—prohibits the use of a direct method of solution. The alternative is to use an iterative method such as the Gauss–Seidel method (see Section 7.3 of Chapter 7). Since the majority of the nodes lie within one material, the general equation for these can easily be specified and repeatedly used in the computer program. The actual writing of the program is made easier by imagining that every FDE is of the same type but requires a different set of coefficients with the different mesh points, a technique introduced in Program 2.5 of Chapter 5, Section 5.2. The program evaluates the coefficients for the entire mesh, based (in the case of temperature-dependent thermal conductivity) on the temperatures at the end of the previous cycle of iteration. This particular procedure will also reduce the amount of computational effort since a given coefficient will usually appear in two FDEs.

With this number of nodes, any feature which will expedite convergence should be employed. Careful choice of the initial node temperatures and over-relaxation (see Section 7.3 of Chapter 7) are two such features. The initial values were those derived in the steady-state 2-D problem discussed in Section 8.9.

8.10.5. Computed Results and Discussion

The computed results are too voluminous to present here. As an illustration, Fig. 8.41 shows the isotherms across a centre-line section through the structure. The distortion of the heat flow pattern due to the presence of the metal pin is clear to see. The effect on the heat transfer rate is most easily calculated by examining the temperature distribution along the outer boundary of the metal shell. Figure 8.42 shows the temperature variation along the outer boundary of the metal shell at a centre-line section. By computing the average surface temperature and comparing this average with the value obtained from the analysis with no pin (as in Section 8.9), the rate of heat transfer is found to increase by about 4%. If the refractory is damaged, resulting in a reduction of thickness between the inner surface and the tip of the fin, the rate of heat transfer is more than doubled since then the tip of the fin may be subjected to a temperature of 1300 °C. It is left to the reader to investigate the effect of pin spacing, etc. Altering the grid size will be the easiest way to model different geometric configurations.

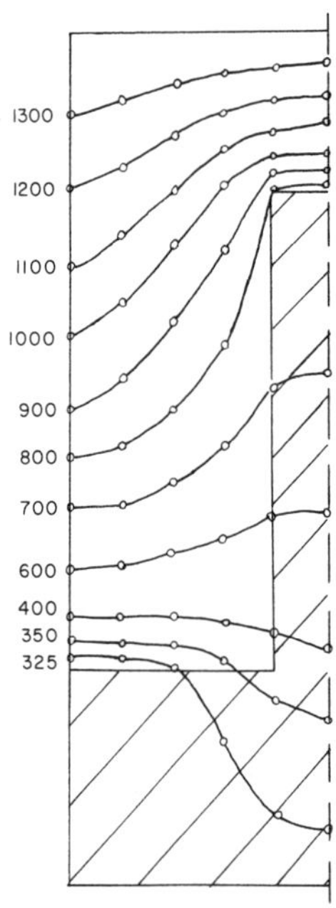

FIG. 8.41. Isotherms surrounding metal pin across centre-line section through the structure.

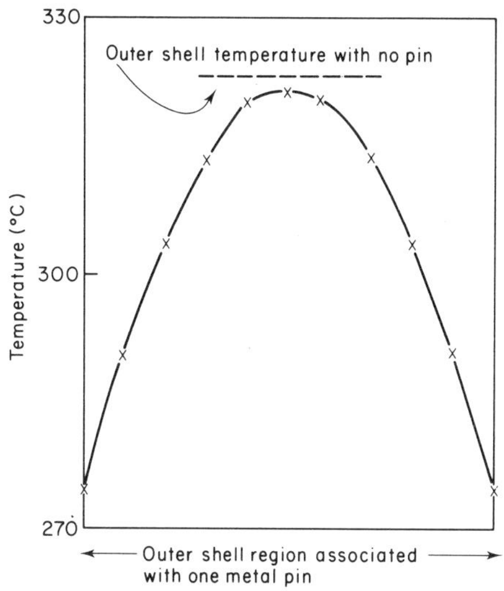

FIG. 8.42. Variation of temperature across outer boundary of metal shell at centre-line section showing the effect of the presence of the metal pin.

The effect of pin size, spacing, length, etc., can be investigated by minor modifications to the work presented here. A simple way to do this, with the same program, is to arrange for a variable size mesh to be accommodated and then merely specifying the grid line positions immediately models different geometric configurations.

BIBLIOGRAPHY

ANON. (1965). *Copper for Busbars*. Copper Development Association, Potters Bar, Herts, UK.

DAVIS, W. B. (1972). Determination of the temperature distribution in a steel slab during quenching. *J. Iron & Steel Inst.*, June.

ECKERT, E. R. G. & DRAKE, R. M. (1959). *Heat and Mass Transfer*, 2nd ed. McGraw-Hill, New York.

ROGERS, L. M. (1971). The application of thermography to plant monitoring and energy conservation. British Steel Corporation (Tubes Division), Report TB/TH/71.

Computer Program Listings

Chapter 5, Program 2.1

```
C       MAIN
C--------PROGRAM 2-1 OF CHAPTER 5
C       SIMPLE STEADY STATE PROBLEM
C       SET INITIAL GUESSES
        T1=0.0
        T2=0.0
        T3=0.0
C       ITERATION LOOP FOR 20 ITERATIONS
        DO 100 L=1,20
        TP1=(100.+100.+T2+200.)/4.
        TP2=(T1+100.+T3+300.)/4.
        TP3=(T2+100.+860.+400.)/4.
C       UPDATE AFTER EACH ITERATION LOOP
        T1=TP1
        T2=TP2
        T3=TP3
C       INTERMEDIATE OUTPUT
        WRITE(5,200)L,T1,T2,T3
  200   FORMAT(4H I =,I3,4HT1 =,F10.3,8H      T2 =,F10.3,8H      T3 =,F10.3)
  100   CONTINUE
        STOP
        END
```

Chapter 5, Program 2.2

```
C       MAIN
C--------PROGRAM 2-2 OF CHAPTER 5
C       SIMPLE STEADY STATE PROBLEM
C       BOUNDARY VALUES VIA A DATA, READ OR EXECUTION STATEMENT
        DATA, TA,TB,TC,TD,TE,TF,TG,TH/
      1   4★100.,860.,400.,300.,200./
C       INITIAL GUESSES
        T1=0.0
        T2=0.0
        T3=0.0
C       WRITE HEADINGS
        WRITE(5,200)
  200   FORMAT(1H1///6X,5HIT NO,5X,2HT1,8X,2HT2,8X,2HT3//)
C       INITIAL OUTPUT
        WRITE(5,201)L,T1,T2,T3
C       ITERATION LOOP
        DO 100 L=1,20
        TP1=(TA+TB+T2+TH)/4.
        TP2=(T1+TC+T3+TG)/4.
        TP3=(T2+TD+TE+TF)/4.
C       UPDATE AFTER EACH ITERATION CYCLE
        T1=TP1
        T2=TP2
        T3=TP3
C       INTERMEDIATE OUTPUT
        WRITE(5,201)L,T1,T2,T3
  201   FORMAT(I10,3F10.3)
  100   CONTINUE
        STOP
        END
```

Chapter 5, Program 2.3

```
C       MAIN
C--------PROGRAM 2-3 OF CHAPTER 5
C       SIMPLE STEADY STATE PROBLEM
C       MORE SOPHISTICATIONS
        DIMENSION T(5,3),TP(5,3),R(5,3),F(5,3)
C       BOUNDARY AND INITIAL VALUES
        DATA T/0.,200.,300.,400.,0.,100.,3★0.,860.,0.,3★100.,0./
        DATA TP/15★0./
C       INITIAL VALUES— SOME ALREADY SET TO ZERO IN DATA
        L=0
        LMAX=100
        IMAX=5
        JMAX=3
        IMM1=IMAX-1
        JMM1=JMAX-1
C       WRITE HEADINGS
        WRITE(5,200)
  200   FORMAT(1H1///6X,5HIT NO,5X,2HT1,8X,2HT2,8X,2HT3//)
C       INITIAL OUTPUT
C       WRITE(5,201)L,T(2,2),T(2,3),T(2,4)
C       ITERATION LOOP
        DO 107 L=1,LMAX
C       CALCULATE TP AT UNKNOWN TEMPERATURE POINTS
        DO 101 I=2,IMM1
        DO 100 J=2,JMM1
        TP(I,J)=(T(I-1,J)+T(I+1,J)+T(I,J-1)+T(I,J+1))/4.
  100   CONTINUE
```

```
  101 CONTINUE
C     CALCULATE FRACTIONAL CHANGES AND UPDATE AFTER EACH ITERATION CYCLE
      DO 103 I=2,IMM1
      DO 102 J=2,JMM1
      R(I,J)=4.★ABS(TP(I,J)-T(I,J))
      F(I,J)=ABS((TP(I,J)-T(I,J))/T(I,J))
      T(I,J)=TP(I,J)
  102 CONTINUE
  103 CONTINUE
C     INTERMEDIATE OUTPUT
      WRITE(5,201)L,T(2,2),T(2,3),T(2,4)
  201 FORMAT(I10,3F10.3)
C     TERMINATION TESTS
      FMAX=0.0
      RES=0.0
      DO 105 I=2,IMM1
      DO 104 J=2,JMM1
      IF(F(I,J).GT.FMAX)FMAX=F(I,J)
      RES=RES+R(I,J)
  104 CONTINUE
  105 CONTINUE
      IF(FMAX.LT.0.005)GO TO 106
      GO TO 107
  106 IF(RES.LT.0.005)GO TO 108
  107 CONTINUE
C     TERMINATION
  108 CONTINUE
      STOP
      END
```

Chapter 5, Program 2.4

```
C        MAIN
C--------PROGRAM 2-4 OF CHAPTER 5
C        SIMPLE STEADY STATE PROBLEM
C        SET INITIAL GUESSES
         T1=0.0
         T2=0.0
         T3=0.0
C        ITERATION LOOP FOR 20 ITERATIONS
         DO 100 L=1,20
         T1=(100.+100.+T2+200.)/4.
         T2=(T1+100.+T3+300.)/4.
         T3=(T2+100.+860.+400.)/4.
C        INTERMEDIATE OUTPUT
         WRITE(5,200)L,T1,T2,T3
  200    FORMAT(4H I =,I3,4HT1 =,F10.3,8H      T2 =,F10.3,8H      T3 =,F10.3)
  100    CONTINUE
         STOP
         END
```

Chapter 5, Program 2.5

```
C        MAIN
C--------PROGRAM 2-5 OF CHAPTER 5
C        SIMPLE STEADY STATE PROBLEM
C        MORE SOPHISTICATIONS
         DIMENSION T(5,3),AN(5,3),AS(5,3),AE(5,3),AW(5,3),AP(5,3)
C        BOUNDARY AND INITIAL VALUES
         DATA T/0.,200.,300.,400.,0.,100.,3★0.,860.,0.,3★100.,0./
         DATA AN,AS,AE,AW/60★1·0/
```

```
C       INITIAL VALUES— SOME ALREADY SET TO ZERO IN DATA
        L=0
        LMAX=100
        IMAX=5
        JMAX=3
        IMM1=IMAX-1
        JMM1=JMAX-1
C       WRITE HEADINGS
        WRITE(5,200)
  200   FORMAT(1H1///6X,5HIT NO,5X,2HT1,8X,2HT2,8X,2HT3//)
C       INITIAL OUTPUT
        WRITE(5,201)L,T(2,2),T(2,3),T(2,4)
C       ITERATION LOOP WITH RESIDUAL CALCULATION
        DO 102 L=1,LMAX
        RES=0.0
        DO 101 I=2,IMM1
        DO 100 J=2,JMM1
        TNEW=AN(I,J)★T(I,J+1)+AS(I,J)★T(I,J-1)
      1   +AE(I,J)★T(I+1,J)+AW(I,J)★T(I-1,J)
        AP(I,J)=AN(I,J)+AS(I,J)+AE(I,J)+AW(I,J)
        TNEW=TNEW/AP(I,J)
        RES=RES+ABS(TNEW-T(I,J))
        T(I,J)=TNEW
  100   CONTINUE
  101   CONTINUE
C       INTERMEDIATE OUTPUT
        WRITE(5,201)L,T(2,2),T(2,3),T(2,4)
  201   FORMAT(I10,3F10.3)
C       TERMINATION TEST
        IF(RES.LT.0.005)GO TO 103
  102   CONTINUE
        TERMINATION
  103   STOP
        END
```

Chapter 6, Program 1.1

```
C        MAIN
C--------PROGRAM 1-1 OF CHAPTER 6
C        SIMPLE TRANSIENT PROBLEM
         DIMENSION T(7),TP(7)
         DATA T/1000.,5★20.,1000./
C        WRITE HEADINGS
         WRITE(5,200)
  200    FORMAT(1H1,5H STEP,12HTEMPERATURES//)
C        ADVANCE IN TIME DELTA T
         DO 100 NDT=1,5
         DO 101 I=2,6
         TP(I)=0.5★(T(I-1)+T(I+1))
  101    CONTINUE
         DO 102 I=2,6
         T(I)=TP(I)
  102    CONTINUE
C        INTERMEDIATE OUTPUT
         WRITE(5,201)NDT,T
  201    FORMAT(I5,7F12.3)
  100    CONTINUE
         STOP
         END
```

Chapter 6, Program 1.2

```
C       MAIN
C-------PROGRAM 1-2 OF CHAPTER 6
C       SIMPLE TRANSIENT PROBLEM
        DIMENSION T(5),TP(5)
        DATA T/510.,4★20./
C       WRITE HEADINGS
        WRITE(5,200)
  200   FORMAT(1H1,5H STEP,12HTEMPERATURES//)
C       ADVANCE IN TIME DELTA T
        DO 100 NDT=1,5
C       RIGHT HAND SYMMETRY CONDITION—
C                    DT/DX = 0 AT T(4)
        T(5)=T(3)
        DO 101 I=2,4
        TP(I)=0.5★(T(I-1)+T(I+1))
  101   CONTINUE
        DO 102 I=2,4
        T(I)=TP(I)
  102   CONTINUE
C       INTERMEDIATE OUTPUT
        WRITE(5,201)
  201   FORMAT(I5,4F12.3)
  100   CONTINUE
        STOP
        END
```

Chapter 7, Program 2.1

```
C   MAIN
C   -----PROGRAM 2-1 OF CHAPTER 7
C   LINEAR EQUATIONS GAUSS ELIMINATION
C   PROGRAM TO SOLVE A SET OF LINEAR EQUATIONS BY GAUSSIAN ELIMINATION.
C   THIS PROGRAM DESIGNED FOR UP TO 40 EQUATIONS.
       DIMENSION A(40,41)
C   READ IN NUMBER OF EQUATIONS, N, AND COEFFICIENTS.
C   COEFFICIENTS ARE PUNCHED 8 PER CARD, EXCEPT EACH EQUATION BEGINS ON A NEW CARD.
       READ(2,100)N
       NP1=N + 1
       DO 10 I = 1,N
    10 READ(2,101)(A(I,J),J=1,NP1)
C   PRINT HEADING
       WRITE(5,200)N
C   ELIMINATE COEFFICIENTS BELOW DIAGONAL.
       DO 20 I = 2,N
       DO 20 J = I,N
C   TEST IF PIVOT ELEMENT IS ZERO. IF SO, SWITCH ROWS.
       IF (A(I-1, I-1)) 1,2,1
     2 IM1 = I- 1
       DO 21 M = I,N
       IF (A(M,IM1)) 3,21,3
     3 DO 22 MM = IM1,NP1
       SAVE = A(M,MM)
       A(M,MM) = A(IM1,MM)
    22 A(IM1,MM) = SAVE
    21 CONTINUE
       WRITE(5,199)
     1 R = A(J,I-1) / A(I-1,I-1)
       DO 20 K = I,NP1
    20 A(J,K) = A(J,K) -R★A(I-1,K)
```

```
C   BACK SUBSTITUTE BY ELIMINATION OF COEFFICIENTS ABOVE DIAGONAL.
        DO 30 I = 2,N
        K = N- I + 2
        R = A(K,NP1) / A(K,K)
        DO 30 J = I,N
        L = N - J + 1
    30 A(L,NP1) = A(L,NP1) -R★A(L,K)
C   SOLUTION IS COLUMN OF CONSTANTS DIVIDED BY CORR DIAG COEFF
        DO 40 I = 1,N
        T = A(I,NP1) / A(I,I)
    40 WRITE(5,201)I,T
        STOP
   100 FORMAT (I2)
   101 FORMAT (8F10.0)
   199 FORMAT (72H COEFFICIENT MATRIX IS SINGULAR. NO UNIQUE SOLUTION TO SET OF
      1  EQUATIONS.)
   200 FORMAT (/ 20H SOLUTION TO SET OF , 13,
      1  34H EQUATIONS BY GAUSSIAN ELIMINATION // )
   201 FORMAT (3H T( , 12, 4H) = , E14.8)
        END

C - - DATA
03
-4.0        1.0        0.0        -400.0
1.0        -4.0        1.0        -400
0.0         1.0       -4.0        -1360.0

C - - END OF DATA

C - - OUTPUT

SOLUTION TO SET OF 3 EQUATIONS BY GAUSSIAN ELIMINATION

T( 1) = 0.16000000E 03
T( 2) = 0.24000000E 03
T( 3) = 0.40000000E 03
```

Chapter 7, Program 2.2

```
C   MAIN
C-----PROGRAM 2-2 OF CHAPTER 7
C   TRIDIAGONAL SYSTEM GAUSS ELIMINATION
C   PROGRAM TO SOLVE TRIDIAGONAL SYSTEM BY ELIMINATION.
C   COEFFICIENTS COMPRESSED INTO AN N BY 4 ARRAY.
C   PROGRAM WILL HANDLE UP TO 500 EQUATIONS. D = DIAGONAL ELEMENTS.
C   A = ABOVE DIAGONAL, B = BELOW DIAGONAL, C = CONSTANT VECTOR.
C   SOLUTION APPEARS IN THE VECTOR C.
        DIMENSION A(500), D(500), B(500), C(500)
C   READ IN NUMBER OF EQUATIONS, N, AND COEFFICIENTS.
C   THE THREE COEFFICIENTS PLUS THE CONSTANT TERM OF EACH EQUATION ARE ON ONE CARD.
        READ(2,100)N,(B(I),D(I),A(I),C(I),I=1,N)
C   PRINT HEADING
        WRITE(5,200)N
C   COMPUTE THE NEW MATRIX. SOLUTION WILL BE STORED IN C ARRAY.
        DO 10 I = 2,N
        R = B(I)/D(I-1)
        D(I) = D(I)- R★A(I-1)
     10 C(I) = C(I)- R★C(I-1)
C   BACK SUBSTITUTION.
        C(N) = C(N)/D(N)
        DO 20 I = 2,N
        J = N - I + 1
     20 C(J) = (C(J)- A(J)★C(J+1))/D(J)
```

```
C   PRINT OUT VALUES OF VARIABLES
      WRITE(5,201)(I,C(I),I=1,N)
      STOP
  100 FORMAT (I3 / (4F10.0))
  200 FORMAT (/ 34H SOLUTION TO TRIDIAGONAL SYSTEM OF , I4, 26H EQUATIONS BY
     1  ELIMINATION //)
  201 FORMAT (3H T( , I3, 4H) = , E14.7)
      END

C - - DATA
005
0.0         4.0       -1.0       0.5
-1.0        4.0       -1.0       1.5
-1.0        4.0       -1.0       2.5
-1.0        4.0       -1.0       3.5
-2.0        4.0        0.0       4.5

C - - END OF DATA

C - - OUTPUT

SOLUTION TO TRIDIAGONAL SYSTEM OF 5 EQUATIONS BY ELIMINATION

T( 1) = 0.3156077E 00
T( 2) = 0.7624309E 00
T( 3) = 0.1234116E 01
T( 4) = 0.1674033E 01
T( 5) = 0.1962017E 01
```

Chapter 7, Program 3.1

```
C   MAIN
C------PROGRAM 3-1 OF CHAPTER 7
C   LINEAR EQUATIONS GAUSS-SEIDEL ITERATION
C   PROGRAM TO SOLVE A SET OF LINEAR EQUATIONS BY GAUSS-SEIDEL ITERATION.
C   PROGRAM HANDLES UP TO 30 EQUATIONS GIVEN IN DOM DIAG FORM TO ENSURE
C      CONVERGENCE.
       DIMENSION A(20,21), T(30)
C   READ IN NO OF EQ, N, LIMIT TO ITERATIONS, ITMAX, TOLERANCE, TOL.
       READ(2,100)N,ITMAX,TOL
C   READ COEFFICIENTS. PUNCH 8 PER CARD, BEGIN EACH EQ ON A NEW CARD.
       NP1 = N + 1
       DO 10 I = 1,N
    10 READ(2,101)(A(I,J),J=1,NP1)
C   READ IN APPROX SOLUTION FOR T, PUNCHED 8 PER CARD.
       READ(2,101) (T(I),I=1,N)
C   PRINT HEADING.
       WRITE(5,200)
C   PERFORM ITERATIONS. STORE MAXIMUM CHANGE IN T VALUES FOR TESTING.
       DO 30 I = 1,ITMAX
       DIFF = 0.0
       DO 20 J = 1,N
       SAVE = T(J)
       T(J) = A(J,NP1)
       DO 21 K = 1,N
       IF (J - K) 22,21,22
    22 T(J) = T(J) - A(J,K)★T(K)
    21 CONTINUE
       T(J) = T(J) /A(J,J)
       IF (DIFF - ABS (T(J) - SAVE)) 23,20,20
    23 DIFF = ABS (T(J) - SAVE)
    20 CONTINUE
```

```
C  INTERMEDIATE WRITES
      WRITE(5,202)I
      WRITE(5,203) (J,T(J),J=1,N)
      IF (TOL - DIFF) 30,40,40
   30 CONTINUE
C  TOLERANCE WAS NOT MET. PRINT NOTE AND LAST VALUES.
      WRITE(5,201)TOL,ITMAX
      GO TO 50
   40 WRITE(5,202)I
   50 WRITE(5,203) (J,T(J),J=1,N)
      STOP
  100 FORMAT (2I3, E14.0)
  101 FORMAT (8F10.0)
  200 FORMAT (/50H SOLUTION TO SET OF LINEAR EQUATIONS BY ITERATION //)
  201 FORMAT(14H TOLERANCE OF , E8.0, 15H NOT MET AFTER ,14, 12H ITERATIONS. / 29H LAST
     1 VALUES CALCULATED WERE /)
  202 FORMAT (16H0T VALUES, AFTER, I4, 17H ITERATIONS, ARE /)
  203 FORMAT (3H T( , I3, 4H) = , E14.7)
      END

C - - DATA
003050            1.E-7
8.0       1.0        -1.0       8.0
1.0       -7.0       2.0        -4.0
2.0       1.0        9.0        12.0

C - - END OF DATA

C - - OUTPUT
```

```
SOLUTION TO SET OF LINEAR EQUATIONS BY ITERATION

T VALUES, AFTER 1 ITERATIONS, ARE

T( 1) = 0.1000000E 01
T( 2) = 0.7142857E 00
T( 3) = 0.1031746E 01

T VALUES, AFTER 2 ITERATIONS, ARE

T( 1) = 0.1039683E 01
T( 2) = 0.1014739E 01
T( 3) = 0.9895440E 00

T VALUES, AFTER 3 ITERATIONS, ARE

T( 1) = 0.9968506E 00
T( 2) = 0.9965626E 00
T( 3) = 0.1001082E 01

T VALUES, AFTER 4 ITERATIONS, ARE

T( 1) = 0.1000565E 01
T( 2) = 0.1000390E 01
T( 3) = 0.9998312E 00

T VALUES, AFTER 5 ITERATIONS, ARE

T( 1) = 0·9999302E 00
T( 2) = 0.9999418E 00
T( 3) = 0.1000022E 01
```

```
T VALUES, AFTER 6 ITERATIONS, ARE

T( 1) = 0.1000010E 01
T( 2) = 0.1000008E 01
T( 3) = 0.9999969E 00

T VALUES, AFTER 7 ITERATIONS, ARE

T( 1) = 0·9999987E 00
T( 2) = 0.9999989E 00
T( 3) = 0.1000000E 01

T VALUES, AFTER 8 ITERATIONS, ARE

T( 1) = 0.1000000E 01
T( 2) = 0.1000000E 01
T( 3) = 0.9999999E 00

T VALUES, AFTER 9 ITERATIONS, ARE

T( 1) = 0.1000000E 01
T( 2) = 0.1000000E 01
T( 3) = 0.1000000E 01

T VALUES, AFTER 10 ITERATIONS, ARE

T( 1) = 0.1000000E 01
T( 2) = 0.1000000E 01
T( 3) = 0.1000000E 01
```

Chapter 7, Program 4.1

```
C   MAIN
C------PROGRAM 4-1 OF CHAPTER 7
C   POISSON EQUATION LIMITED STORAGE GAUSS-SEIDEL
C   OVER-RELAXATION
C   PROGRAM FOR SOLVING POISSON EQUATION, COND★DEL SQ T + H = 0 BY G-S SOR METHOD.
        DIMENSION T(20,20)
C   READ IN PARAMETERS, PUNCH HEADING.
        READ(2,99) NWIDE, NHIGH,COND,TOL,H,DELX
        XK = H/COND
        ALPHA=0.5-3.14159/2.8284★SQRT(1./NWIDE★★2+1./NHIGH★★2)
        WRITE(5,100) XK, ALPHA
        NWPI = NWIDE + 1
        NHPI = NHIGH + 1
C   BOUNDARY VALUES
        DO 10 I = 1,NHPI, NHIGH
     10 READ(2,101) (T(I,J),J=1,NWPI)
        DO 11 J = 1, NWPI, NWIDE
     11 READ(2,101) (T(I,J),I=2,NHIGH )
C   INITIALIZE INTERIOR POINTS
        DO 12 I = 2,NHIGH
        DO 12 J = 2,NWIDE
     12 T(I,J) = 0.0
```

```
C   COMPUTE VALUES BY G-S SOR METHOD. END WHEN MAX CHANGE IS LESS THAN THE
C       TOLERANCE.
        DO 14 K = 1,50
        DEVMX = 0.0
        DO 13 I = 2,NHIGH
        DO 13 J = 2,NWIDE
        SAVE = T(I,J)
        T(I,J) = ALPHA★(T(I+1,J) + T(I-1,J) + T(I,J+1) + T(I,J-1) + XK★DELX★★2 - (4. -
       1    1./ALPHA)★T(I,J))
        DEV = ABS(T(I,J)-SAVE)
        IF (DEV- DEVMX) 13,13,15
     15 DEVMX = DEV
     13 CONTINUE
        IF (TOL- DEVMX) 14,16,16
     14 CONTINUE
        WRITE(5,102)
        GO TO 17
     16 WRITE(5,104) K
     17 DO 30 I = 1,NHPI
     30 WRITE(5,103) I, (T(I,J), J=1,NWPI)
C   HEAT FLOW BY CONDUCTING ROD ANALOGY.
        TOTAL = 0.
        DO 40 I = 2,NHIGH
     40 TOTAL = T(I,2) - T(I,1) + TOTAL
        DO 50 I = 2,NHIGH
     50 TOTAL = TOTAL + T(I,NWIDE) - T(I,NWPI)
        DO 60 J = 2,NWIDE
     60 TOTAL = TOTAL + T(2,J) - T(1,J)
        DO 70 J = 2,NWIDE
     70 TOTAL = TOTAL + T(NHIGH,J) - T(NHPI,J)
        TOTAL = TOTAL★COND
        WRITE(5,105) TOTAL
```

```
C   INTERNAL HEAT GENERATION.
      VOL = NWIDE★NHIGH★DELX★★2★1.
      Q = VOL★H
      WRITE(5,106) Q
      STOP
   99 FORMAT(2I2/5F10.3)
  100 FORMAT(/71H SOLUTION TO POISSON EQTN - COND★ DEL SQ T + H =0 - BY G-S SOR
     1  METHOD./ 13H FOR H/COND =,F8.1,29H, USING RELAXATION FACTOR OF, F5.3)
  101 FORMAT (8F10.0)
  102 FORMAT (//51H TOLERANCE TO END COMPUTATION NOT MET IN 50 CYCLES./32H LAST
     1  ROUND OF CALCULATIONS IS -)
  103 FORMAT (4H0R0W, I2/(1H ,16F7.3))
  104 FORMAT (// I3, 21H ITERATIONS WERE USED)
  105 FORMAT(//39H HEAT FLOW BY CONDUCTING ROD ANALOGY = ,F10.3,7H WATTS.)
  106 FORMAT(/28H INTERNAL HEAT GENERATION = ,F10.3,7H WATTS.)
      END

C - - DATA
0816
    40.000     0.001  800000.000      0.005
        0.         0.         0.         0.        0.        0.        0.
        0.
        0.         0.         0.         0.        0.        0.        0.
        0.
        0.         0.         0.         0.        0.        0.        0.
        0.         0.         0.         0.        0.        0.        0.
        0.         0.         0.         0.        0.        0.        0.
        0.         0.         0.         0.        0.        0.        0.

C - - END OF DATA

C - - OUTPUT
```

```
SOLUTION TO POISSON EQTN— COND★ DEL SQ T + H =0— BY G−S SOR METHOD.
FOR H/COND = 20000.0, USING RELAXATION FACTOR OF 0.345

31 ITERATIONS WERE USED

ROW 1
   0.000  0.000  0.000  0.000  0.000  0.000  0.000  0.000  0.000

ROW 2
   0.000  0.620  0.982  1.176  1.238  1.177  0.982  0.620  0.000

ROW 3
   0.000  0.998  1.632  1.986  2.100  1.986  1.633  0.998  0.000

ROW 4
   0.000  1.240  2.064  2.536  2.689  2.536  2.065  1.240  0.000

ROW 5
   0.000  1.397  2.350  2.904  3.086  2.905  2.351  1.398  0.000

ROW 6
   0.000  1.498  2.536  3.146  3.347  3.146  2.538  1.499  0.000

ROW 7
   0.000  1.561  2.652  3.296  3.509  3.297  2.653  1.562  0.000

ROW 8
   0.000  1.596  2.715  3.379  3.598  3.379  2.716  1.597  0.000
```

```
ROW 9
  0.000  1.607  2.735  3.405  3.627  3.405  2.736  1.607  0.000

ROW10
  0.000  1.596  2.716  3.379  3.599  3.380  2.717  1.597  0.000

ROW11
  0.000  1.562  2.653  3.298  3.511  3.298  2.654  1.563  0.000

ROW12
  0.000  1.499  2.538  3.148  3.349  3.148  2.539  1.500  0.000

ROW13
  0.000  1.398  2.352  2.906  3.088  2.907  2.352  1.398  0.000

ROW14
  0.000  1.241  2.066  2.538  2.691  2.538  2.066  1.241  0.000

ROW15
  0.000  0.999  1.634  1.988  2.101  1.988  1.634  0.999  0.000

ROW16
  0.000  0.620  0.983  1.177  1.239  1.177  0.983  0.620  0.000

ROW17
  0.000  0.000  0.000  0.000  0.000  0.000  0.000  0.000  0.000

HEAT FLOW BY CONDUCTING ROD ANALOGY =  2098.668 WATTS.

INTERNAL HEAT GENERATION =  2560.000 WATTS.
```

Chapter 7, Program 5.1

```
C   MAIN
C-----PROGRAM 5-1 OF CHAPTER 7
C   PARABOLIC EXPLICIT DIRICHLET
C   PROGRAM TO COMPUTE THE TEMPERATURE DISTRIBUTION AS A FUNCTION OF TIME IN A ONE
C      DIMENSIONAL HEAT FLOW SYSTEM WITH CONSTANT END CONDITIONS, USING THE EXPLICIT
C      METHOD.
       DIMENSION T(2,50)
C   READ IN SYSTEM PARAMETERS. EL = TOTAL LENGTH, COND = THERMAL CONDUCTIVITY,SPHT
C      = SPECIFIC HEAT, DENS = DENSITY, DELX = DISTANCE INTERVAL.
C   ALL MUST BE IN CONSISTENT UNITS.
C   ALSO READ FOURIER NUMBER = (COND★DTIME)/(SPHT★DENS★DELX★★2), AND FINAL TIME
C      TO WHICH COMPUTATIONS WILL CONTINUE.
       READ(2,100) EL, COND, SPHT, DENS, DELX, FO, FINAL
C   READ IN INITIAL TEMPERATURES AFTER FIRST COMPUTING NUMBER OF T VALUES IN THE
C      ARRAY.
       NUMB = EL/DELX+ 1.
       READ(2,100)(T(1,I) , I = 1, NUMB)
C   READ IN BOUNDARY TEMPERATURES.
       READ(2,100) TLHE,TRHE
       T(1,1) = (TLHE+ T(1,1))/2.
       T(1,NUMB) = (TRHE+ T(1,NUMB))/2.
C   PRINT HEADING AND INITIAL TEMPERATURES.
       TIME=0.
       WRITE(5,200)DELX
       WRITE(5,201)TIME,(T(1,I),I=1,NUMB)
```

```
C   COMPUTE SET OF TEMPERATURES AFTER ONE DTIME INTERVAL.
        DTIME = FO★SPHT★DENS★(DELX★★2)/COND
        NM1 = NUMB - 1
        TIME = TIME + DTIME
        DO 10 I = 2,NM1
   10   T(2,I) = (1.-2.★FO)★T(1,I) +FO★(T(1,I+1)+T(1,I-1))
        T(1,1) = TLHE
        T(1,NUMB) = TRHE
        GO TO 15
    5   TIME = TIME + DTIME
        DO 11 I = 2,NM1
   11   T(2,I) = (1.-2.★FO)★T(1,I) +FO★(T(1,I+1)+T(1,I-1))
   15   CONTINUE
C   SET NEW TEMPERATURES INTO OLD ARRAY AND PRINT.
        DO 20 I = 2,NM1
   20   T(1,I) = T(2,I)
        WRITE(5,201)TIME,(T(1,I),I=1,NUMB)
        IF(TIME— FINAL) 5,30,30
   30   CONTINUE
        STOP
  100   FORMAT(8F10.3)
  200   FORMAT (1H1, 30X, 60H UNSTEADY STATE TEMPERATURE DISTRIBUTION BY EXPLICIT
      1   METHOD /31X, 40H TEMPERATURES ARE GIVEN AT SPACING OF F6.3, 10H M. APART )
  201   FORMAT (1HO, 11H AT TIME = F10.3, 8H SECONDS /(1H , 10F10.3))
        END

C - - DATA
     0.100     0.210     0.920  2700.000      0.02      0.25     100.0
    20.000    20.000    20.000    20.000    20.000    20.000
     0.000   100.000

C - - END OF DATA

C - - OUTPUT
```

```
UNSTEADY STATE TEMPERATURE DISTRIBUTION BY EXPLICIT METHOD
TEMPERATURES ARE GIVEN AT SPACING OF     0.020   M. APART

AT TIME = 0.000 SECONDS
    10.000   20.000   20.000   20.000   20.000    60.000

AT TIME = 1.183 SECONDS
     0.000   17.500   20.000   20.000   30.000   100.000

AT TIME = 2.366 SECONDS
     0.000   13.750   19.375   22.500   45.000   100.000

AT TIME = 3.549 SECONDS
     0.000   11.719   18.750   27.344   53.125   100.000

AT TIME = 4.731 SECONDS
     0.000   10.547   19.141   31.641   58.398   100.000

AT TIME = 5.914 SECONDS
     0.000   10.059   20.117   35.205   62.109   100.000

AT TIME = 7.097 SECONDS
     0.000   10.059   21.375   38.159   64.856   100.000

AT TIME = 8.280 SECONDS
     0.000   10.373   22.742   40.637   66.968   100.000

AT TIME = 30.754 SECONDS
     0.000   18.319   37.280   57.279   78.318   100.000
```

```
AT TIME =  31.937 SECONDS
          0.000     18.479     37.539     57.539     78.479    100.000

AT TIME =  33.120 SECONDS
          0.000     18.625     37.774     57.774     78.624    100.000

AT TIME =  34.303 SECONDS
          0.000     18.756     37.987     57.987     78.756    100.000

AT TIME =  96.994 SECONDS
          0.000     19.994     39.990     59.990     79.994    100.000

AT TIME =  98.177 SECONDS
          0.000     19.994     39.991     59.991     79.994    100.000

AT TIME =  99.360 SECONDS
          0.000     19.995     39.992     59.992     79.995    100.000

AT TIME = 100.543 SECONDS
          0.000     19.995     39.993     59.993     79.995    100.000
```

Chapter 7, Program 5.2

```
C   MAIN
C------PROGRAM 5-2 OF CHAPTER 7
C   PARABOLIC CRANK-NICOLSON NEUMANN
C   THIS PROGRAM DETERMINES TEMPERATURES AS A FUNCTION OF TIME IN A ONE-DIMENSIONAL
C       SYSTEM LOSING HEAT FROM BOTH ENDS ACCORDING TO
C          DT/DX = COEFF★(T - TAMB)
C   WHERE    COEFF = (HEAT TRANSFER COEFFICIENT/SOLID CONDUCTIVITY) AND TAMB IS THE
C       CONSTANT AMBIENT TEMPERATURE.
C   A SUBROUTINE IS EMPLOYED TO SOLVE THE SET OF EQUATIONS WHICH RESULT FROM
C       APPLYING THE CRANK-NICOLSON METHOD.
C   THE TEMPERATURES ARE OUTPUT ABOUT EVERY 100 SECONDS UNTIL THE MIDPOINT
C       TEMPERATURE DROPS TO APPROXIMATELY 60 DEGREES CENTIGRADE.
        DIMENSION T(50,4), TINIT(50)
        READ(2,100) SPHT, DENS, COND, HCOEF, TAMB, EL, FO, NOINT
C     READ INITIAL VALUES OF TEMPERATURES IN MATERIAL
        N = NOINT +1
        READ(2,101) (TINIT(I),I=1,N)
C   COMPUTE INITIAL VALUES OF DELTA X, AND DELTA TIME
        TIME = 0.
        TIMEP = 100.
        XINT = NOINT
        DELX = EL/XINT
        DTIME = FO★DENS★SPHT★DELX★★2/COND
        WRITE(5,199)
        WRITE(5,200) TIME,(TINIT(I),I = 1,N)
```

```
C   GET COEFFICIENTS OF SYSTEM
        COEFF = HCOEF/COND
    21  T(1,1) = 0.
        T(1,2) = -4. -2.★COEFF★DELX
        T(1,3) = 2.
        T(N,1) = 2.
        T(N,2) = T(1,2)
        T(N,3) = 0.
        DO 20 I = 2,NOINT
        T(I,1) = 1.
        T(I,2) = -4.
    20  T(I,3) = 1.
        T(1,4) = -2.★TINIT(2)+ 2.★COEFF★DELX★TINIT(1)-4.★COEFF★DELX★TAMB
        T(N,4) = -2.★TINIT(N-1)+2.★COEFF★DELX★TINIT(N)-4.★COEFF★DELX★TAMB
        DO 30 I = 2,NOINT
    30  T(I,4) = -TINIT (I-1) - TINIT (I+1)
C   COMPUTES VALUES. PRINT OUT ABOUT EVERY 100 SECONDS.
        TIME = TIME + DTIME
        CALL TRDG (T,N)
C   PUT NEW VALUES INTO TINIT AND SEE IF TIME TO PRINT OUT.
        DO 40 I = 1,N
    40  TINIT(I) = T(I,4)
        IF(TIME -TIMEP) 21,22,22
    22  WRITE(5,200)TIME,(T(I,4), I=1,N)
        TIMEP = TIMEP + 100.
        IF(TINIT(11)-60.) 23,21,21
    23  CONTINUE
        STOP
   100  FORMAT(7F9.2,I5)
   101  FORMAT(8F10.2)
   199  FORMAT(/57H1SOLUTION TO HEAT FLOW PROBLEM BY CRANK-NICOLSON METHOD/)
   200  FORMAT (/5H T = F6.1, 8H SECONDS / (1H 10F8.2))
        END
```

```
      SUBROUTINE TRDG (X,N)
C   THIS SUBROUTINE SOLVES A TRIDIAGONAL SYSTEM OF EQ BY DIRECT GAUSS ELIMINATION,
C     LIKE PROGRAM 2.2 OF CHAPTER 7
      DIMENSION X(50,4),A(50),B(50),C(50),D(50)
      DO 30 I = 1,N
      B(I) = X(I,1)
      D(I) = X(I,2)
      A(I) = X(I,3)
   30 C(I) = X(I,4)

      DO 10 I = 2,N
      R = B(I)/D(I-1)
      D(I) = D(I) - R★A(I-1)
   10 C(I) = C(I) - R★C(I-1)
C   BACK SUBSTITUTION.
      C(N) = C(N)/D(N)
      DO 20 I = 2,N
      J = N - I + 1
   20 C(J) = (C(J) - A(J)★C(J+1))/D(J)

      DO 40 I = 1,N
   40 X(I,4) = C(I)
      RETURN
      END

C - - DATA
  495.00  8892.00   400.00  1600.00    20.00     1.00     1.00   20
    50.00    52.50    55.00    57.50    60.00    62.50    65.00
    70.00    72.50    75.00    77.50    80.00    82.50    85.00
    90.00    92.50    95.00    97.50   100.00

C - - END OF DATA

C - - OUTPUT
```

```
SOLUTION TO HEAT FLOW PROBLEM BY CRANK–NICOLSON METHOD
T =    0.0 SECONDS
   50.00  52.50  55.00  57.50  60.00  62.50  65.00  67.50  70.00  72.50
   75.00  77.50  80.00  82.50  85.00  87.50  90.00  92.50  95.00  97.50
  100.00

T =  110.0 SECONDS
   44.26  48.84  52.87  56.36  59.44  62.25  64.89  67.46  69.98  72.49
   74.99  77.47  79.91  82.28  84.44  86.17  87.04  86.45  83.72  78.14
   69.66

T =  220.1 SECONDS
   42.72  47.11  51.18  54.90  58.31  61.45  64.37  67.12  69.75  72.28
   74.69  76.96  78.99  80.64  81.72  81.97  81.09  78.78  74.80  69.03
   61.49

T =  907.8 SECONDS
   38.72  42.39  45.90  49.21  52.29  55.12  57.65  59.85  61.67  63.09
   64.05  64.52  64.46  63.85  62.66  60.87  58.49  55.52  51.99  47.95
   43.45

T = 1017.9 SECONDS
   38.15  41.71  45.10  48.29  51.24  53.92  56.31  58.35  60.03  61.29
   62.11  62.45  62.29  61.60  60.37  58.59  56.26  53.40  50.04  46.22
   42.00

T = 1100.4 SECONDS
   37.72  41.19  44.49  47.58  50.43  53.02  55.30  57.24  58.81  59.97
   60.70  60.97  60.74  60.01  58.77  57.00  54.73  51.95  48.71  45.05
   41.01

T = 1210.4 SECONDS
   37.13  40.49  43.66  46.62  49.35  51.80  53.95  55.76  57.20  58.25
   58.88  59.06  58.77  58.01  56.76  55.03  52.83  50.17  47.09  43.61
   39.80
```

Index